Mobile App Marketing And Monetization

by Alex Genadinik

DEDICATION

Dedicated to my mother and grandmother who are the biggest entrepreneurs I know.

CONTENTS

Chapter 1: Examining The Mobile App As A Business   **10**

i. Welcome - read this first   10

ii. My Story   10

iii. Creating The App   12

iv. Brief Mobile App Marketing Introduction   13

v. Brief Mobile App Monetization Introduction   14

vi. Setting Reasonable Goals For Your App   15

vii. On Which Platform Should You Release Your App First?   15

viii. Basic Business Planning Questions To Ask Yourself Before Beginning Work On An App   19

Chapter 2: Strategies To Promote A Mobile App   **21**

i. App Store Search And ASO   22

ii. Publicity And Press Coverage For Your App   33

iii. Social Sharing From Social Networks   40

iv. Social Sharing And User Invites Generated From Inside Your App 41

v. Website Promotion   41

vi. Cross Promote And Partner With Other Apps And Websites   43

vii. Pay For Advertising   43

viii. Email Marketing And Growing Your Email List   44

iv. Speak At Events   44

v. Create A YouTube Channel   45

vi. Keep Your App Freemium   46

vii. Google Search (SEO) For Your App Store URL   47

viii. Establish A Presence On Any Large Platform   48

xiv. Become A Great Marketer   48

xv. If Your App Is Android, Release It On Multiple App Stores   49

xvi. Participate In Forums   49

xvii. How To Ensure That You Get Good App Reviews   50

xviii. How To Avoid Bad Reviews In The App Stores   52

xix. How To Boost Social And Engagement Signals Of Your App   55

xx. 3 Ways To Launch Your App   61

xxi. Translate Your App's Title And Description To Other Languages 66

Chapter 3: How To Monetize Your Mobile App   68

i. Should A Mobile App Be Free Or Paid?   69

ii. How Much To Charge For An App If You Do Make It Paid   69

iii. Making Money With Ads   72

iv. Making Money With Affiliate Products   75

v. Making Money With In-App Purchases   75

vi. Making Money By Selling Your Own Products   76

vii. Making Money By Selling Services   77

viii. The Whale Monetization Pattern   78

ix. How To Maximize Your App's Revenue   80

Chapter 4: Additional Tactics And Strategies   83

i. Comparing Mobile App Stores: GooglePlay vs. Apple App Store vs. Kindle vs. Windows Phone Store vs. Blackberry vs. NOOK   83

ii. Should You Develop Your Apps Natively?   86

iii. General Features To Add To Your App That Will Help It Grow And Make Money   88

iv. What Kinds Of Apps Get An Investment?   91

v. Crowdfunding For Apps   94

vi. 10 Fundraising Strategies   96

vii. Dive Into Mobile App Business Models   99

viii. Mobile App Exit Strategies   104

iv. How To Get Featured In App Stores   106

Chapter 5: Common Types Of Apps And Strategies For Them   **109**

i. Game Apps   109

ii. Social And Photo Apps   112

iii. Dating Apps   114

Chapter 6: Case Study Of ASO For My Apps   **116**

Chapter 1: Examining The Mobile App As A Business

i. Welcome - read this first

Hello, I am excited that you got this book, and I want to extend a very warm welcome to you. In writing this book I did my best to cover every element that will help you succeed with your mobile app business.

GIFTS FOR YOU

In an effort to give you extra value from this book, at the end of the book there are 3 additional resources that I made free for you. They include free personal feedback and advice about your app, free extra business training and more.

I hope you enjoy the book, and I hope to hear from you when you complete it. Whether you like the book or not, I'd love to hear your thoughts on it. My personal email address is: alex.genadinik@gmail.com

I look forward to hearing from you after you complete the book.

ii. My Story

Let me tell you a little bit about me, and why I am qualified to write on this topic. I created my first app in early 2012. Since then I created 8 Android apps for business, 6 iOS apps, and 4 Amazon Kindle apps. The apps have passed 1,000,000 downloads (many of them paid). The apps earn me a full-time living, and have helped me expand my business beyond the app world. The apps I built are business apps that help entrepreneurs plan, start and grow their businesses. They are very loved apps. In fact, if you search for the word "business" on Android, my app is #1 and has been at #1 for over 2 years as of this latest revision of December 2015.

Update on January, 2017: In the last 6 months, my business plan app is still near the top of the search "business" on Android, but has jumped between the 1st and 4th spot.

It took me a few months to reach that ranking, but after I achieved it, I've dominated the #1 spot since late 2012, and in this book I'll share all the tips and strategies I used to get my apps to rank so highly, and reach 1,000,000 downloads - so you can do it too. In fact, **my goal is that you would get much more than 1,000,000 downloads**.

In this book you will learn how to plan a winning app before you even begin creating it, how to promote a mobile app like a pro, and how to make money from your app.

With all that said, I want to caution you to keep your initial expectations reasonable. I certainly was not an overnight success. I had to scratch and claw for every small win. My success came only an incredible amount of hard work that I sustained over a long period of time. Although the sky's the limit, the app business isn't an easy business. Competition is fierce, learning curves are steep, monetization strategies don't always work as well we you would like them to work, learning to program and design apps isn't simple, and development time is expensive.

For these reasons I want to urge you to be prepared to work hard over the long-term. Daydreaming about success is fun, but you will have to put serious sweat and hard work into your app business to make it a true success. Get ready!

iii. Creating The App

There are two ways to create a mobile app. You can program the app by yourself like I did, or you can hire a developer or an agency to build your app. Both of these methods have their pros and cons.

If you are a developer, creating your app is much cheaper and faster. The challenge is to continue improving your Android and iPhone development skills, making ongoing app updates, keeping up with latest operating system updates, and changes in programming languages. Creating and maintaining your apps is a difficult and stressful experience which leaves you little time to promote the app and work on business tasks that the app requires.

If you are not a developer, before you go out and risk thousands or tens of thousands of dollars by hiring someone or some agency to create your app, I want to encourage you to be open minded about learning how to develop apps. If you learn to create apps on your own, you will learn a very marketable skill, and you will save tens of thousands of dollars because you won't have to hire a developer. And when you will have other app ideas in the future, you will be able to simply create them on your own, without needing outside help.

If programming a mobile app is not for you, you can hire a mobile app developer on a site like upwork.com, or you can hire any of the mobile app development agencies. As a rule of thumb, the more you pay, the higher level of quality you will get. It isn't unusual for an app to cost tens of thousands of dollars when it is all said and done.

One common pitfall when hiring a developer or a mobile app development agency is that most entrepreneurs just want the first version of their app developed, and spend most of their funds on launching the first version of their apps. But the truth is that while the launch of an app might seem like a huge deal, the first version of the app is just the beginning. The key to success is to continue improving the app by constantly creating new updates, and making the app better and better over time. Those updates cost money. Over time, this can cost well over $100,000 in addition to the cost to develop the initial version of the app. That is very risky when you consider that over 99% of mobile apps do not go on to raise money from professional investors, and do not make the money back for the entrepreneur.

iv. Brief Mobile App Marketing Introduction

How you promote your app should depend on the kind of app you have, and the level of competition in your app's niche. For most apps, the biggest source of downloads is app store search which is achieved by a process called ASO (App Store Optimization). The second biggest source of downloads is social sharing. After that, the next large potential sources of downloads are publicity and being featured in the app stores (Apple App Store and Android's Google Play). This book will cover all these strategies in more detail in a chapter that will focus on marketing for your mobile apps.

v. Brief Mobile App Monetization Introduction

Mobile app monetization is probably the single most challenging part of your mobile app business. For my apps, finding the right ways in which my apps would make money was the most time consuming, difficult and mind boggling part of the process.

The challenge wasn't figuring out which monetization strategies I was able to try. The monetization strategies for apps are relatively common knowledge. The real challenge was making those strategies effective at bringing in enough income on a per-user basis.

The most common way to monetize apps is by publishing ads on the apps. The second most common way to make money from your apps is to enable in-app purchases (subscriptions, consumable purchases and non-consumable purchases). Other ways to make money from your apps is to sell your own products via the apps, or to sell other products as an affiliate.

This book will cover these and additional monetization strategies in much more detail shortly.

vi. Setting Reasonable Goals For Your App

You should also set realistic goals for your app. I realize all entrepreneurs want to sell their business for a billion dollars (or more), but that isn't the reality for most apps. Many apps have very limited potential either due to poor discoverability and natural marketing options, low demand, limited social sharing potential, or being in a niche that is too competitive.

Think through the potential of your app. This will give you a sense of how much time and money it may be reasonable to invest into your app. Doing this will help you identify the right opportunity for you.

If you are in the process of creating your first app, I realize that it is difficult to create a good business plan and strategy for it without having prior experience. For that reason, as one of the free gifts I am offering you at the end of the book is an opportunity to get my feedback on your app strategy.

vii. On Which Platform Should You Release Your App First?

The mobile app world first truly blossomed on the iOS platform. Despite Android quickly gaining ground on iOS, for a number of years there was no question on which platform developers should release their apps on first. It was iOS because it had the most early adopters, overall users, better monetization rates, the devices were better, and there was an aura of coolness associated with iOS.

In recent years, Android fully caught up to iOS, surpassed it in total devices on which the operating system is installed, and many early iOS adopters switched to Android. The decision of where to first release your app has become much less clear.

In the case of my apps, I first launched them on Android simply because I owned an Android device at that time, and it was easier for me to test the apps on the Android device that I owned as I developed them. I made my original decision based on that coincidence, but I learned many of Android's advantages along the way. This kind of decision-making may seem too simplistic, and it was. But it was very practical, less financially risky, and more affordable for me as someone who was just entering and experimenting in the app world. It made my barrier to entry much smaller.

It goes to show that there are very basic criteria on which it is OK to base your decision. Now let's get into more complex decision criteria.

The biggest development advantage that Android has over iOS is that Android has absolutely no review process. You can update your app multiple times a day if you want. Since there is no review process, your app can be as bad as you want it to be when you first release it to the Google Play store. Of course, no one wants their app to be bad, but this nuance allowed me to release an extremely "beta" version of the app very early in my development. I wasn't too worried about getting bad reviews because if I had gotten bad reviews, I could always just take the app off the app store, improve it, and release it again as a new app without those bad reviews. The early bad reviews wouldn't hurt the app, so I didn't have to worry about that.

The advantage of releasing early and being able to update the app whenever I wanted to proved to be tremendous. On a typical day, I would wake up, check my analytics to learn how people used the app the night before, based on that usability data I would figure out the next strategy, update one or two features on the app, and release a new version of the app that evening or afternoon. Next morning I would again check my analytics to observe how users used this new version of the app, and make a new set of decisions from monitoring how users used the app, and make appropriate improvements again. Sometimes I was able to do this cycle as often as a few times per day if the app changes were simple.

As you can imagine, being able to have such frequent app update cycles allowed me to improve the app very rapidly, and quickly get my app to a point where users really liked the app. This flexibility of the Android platform allowed me to take my very low quality early "beta" version of the app and turn it into a very liked app in a matter of a month or two.

That kind of a rapid development and app improvement cycle is much more difficult to do on iOS. Whereas my development cycle on Android was sometimes as short as half a day or a day, on iOS it was averaging about a week or slightly less than that because of the lengthy app review process that is a part of releasing app updates in the Apple App Store. On Android I was able to improve the app and experiment many times faster than on iOS, and that made all the difference.

I ended up using Android as a platform where I would innovate, and iOS as a place where I would only add features which I perfected and tested on Android. Once I would get my Android app's features to a good place where users liked the new features, I would take all the features enjoyed by my Android users that took me many experiments to hone in on, and add it to the iOS app in a monthly or bi-weekly update.

The speed of improvement of my app on Android was equivalent to a sports car while the speed of app improvement on iOS was comparable to a slow moving horse and buggy.

If your goal is to rapidly improve your app using the "Lean Start-up methodology" where you release an early version of the app, observe results, make improvements based on the observations, and release the app back to be used by users, and continue this cycle on an ongoing basis, Android is a great way to go.

On the other hand, if you want to have a big launch with pizazz and publicity, iOS is probably still the cooler platform that is more hip. Many developers I know still release apps for iOS first, and think about Android second. Additionally, if you have an app that uses hardware like the camera, it can be much simpler to develop apps for iOS because there are literally thousands of different kinds of Android devices which different cameras on all of them. Those devices have very different hardware, and it can sometimes be a nightmare to develop apps for hardware that can be so varied. In such cases, iOS offers a more reasonable platform.

Lastly, iOS is still the place that has the more affluent user base. Chances are that you will make more money on iOS on a per-user basis.

Today, you often don't have to make to choose between Android and iOS, and can launch on both platforms at the same time. Development costs have come down and many technologies offer the ability to launch on Android and iOS simultaneously. If it isn't too much extra cost, time or effort, the ideal launch is a simultaneous one on both platforms.

viii. Basic Business Planning Questions To Ask Yourself Before Beginning Work On An App

While you don't necessarily need a formal business plan, it is a great idea to do some basic business planning for your app before you begin work on your app. The purpose of this early business planning is to catch potential mistakes during an early stage of your app business.

The earlier you catch a mistake the cheaper it is to fix that mistake, and adjust the rest of your business strategy to avoid it.

When you are considering an app idea, try to think about how that particular app can be promoted, developed, how you can make money from it, and where to take the app long-term.

The more rigor you apply to answering these questions, the stronger your app strategy will be.

If you are new to the app world, and planning your first app, it might be difficult to answer these questions correctly due to lack of experience in the industry. For that reason, as I mentioned earlier, you are welcome to get free feedback from me about your app. At the end of this book you will find all the extra ways in which I go out of my way to help you. Providing feedback on your app ideas is one of those ways, and you are welcome to take me up on my offer.

Chapter 2: Strategies To Promote A Mobile App

For an app to succeed, you will need to generate hundreds of thousands, if not millions or tens of millions of downloads. In this chapter I will present over 20 mobile app marketing strategies to get the downloads you need for your app.

The strategies in this chapter are some of the most effective and common in the mobile app industry. They are used by some of the biggest apps out there. These are also the strategies that have been effective for me to promote my apps to generate a cumulative download total of well over 1,000,000 and become some of the top apps used by entrepreneurs to plan, start and grow their businesses. I am just a single developer, probably just like you are. I am not a multi-million dollar app development studio. If I was able to get the strategies presented here to work for me, so can you.

What I recommend to my coaching clients is that they should have a marketing strategy created before they decide on a whether working on an app is a good idea. A big part of the business planning for an app is whether it can be promoted well, and grow.

i. App Store Search And ASO

The app stores are the number one source of app downloads for the vast majority of apps, and the bulk of those downloads come from app store search and apps being featured. Since most apps can more reliably get downloads from search, let's focus on search first, and let's get back to being featured in the Apple App Store and Google Play Store later in this chapter.

The process of making your app easily discovered by app store search is called ASO (app store optimization).

There are three steps you need to take in order to optimize your app store listing to get the largest number of downloads possible. I'll give a brief overview of these three steps, and then we'll dive into them a little deeper.

The first step is to choose the right keywords for which you will try to get your app to rank when people search in the app stores. Choosing the right keywords is very important. They are the foundation of your search strategy. If you pick the wrong keywords, you will get unengaged and disinterested traffic. And if you pick keywords that are too competitive to rank for, you simply won't be able to rank and won't get any traffic from search, and will miss out on the amazing opportunity to get a significant amount of downloads from people who find your app through by searching.

If you are not experienced with keyword research, you can try to get an SEO or ASO professional to help you understand what keywords to try to rank for in the app stores, the amount of competition per keyword, and the volume of demand for those keywords.

Once you understand what keywords you want your app to rank for, the next step is to optimize your app's app store listing to maximize the conversion rate of people choosing it when browsing other apps listed in search results, and proceeding to download it.

To optimize your app store listing for download conversion, create a professional app icon (preferably a bright one that stands out), beautiful screenshots, and write an amazing description and title which are at the same time rich with the keywords you are targeting for search, and also very appealing to people who will be reading that description when they will be deciding whether to download your app. You want to get people curious and excited about downloading your app.

The third step towards perfect app store optimization has almost nothing to do with the actual app store. It has to do with your marketing and how well your app is made. Once people get your app, make sure the app is engaging, and people actually like it and use it. Work to increase the amount of time people spend using the app, and the number of times people open the app. Those engagement signals are search engine ranking factors that will help your app rank higher in the app store rankings for the keywords you chose in step one.

The last part of step three is to make sure your app gets good reviews. Reviews are a big part of your app store listing. So first and foremost, your app must satisfy and delight your users. Otherwise, everything else will be much more difficult. Good reviews will help to convince new potential users to download your app and will also help to boost you in the app stores while bad reviews will deter people from downloading your app.

If you do each of the above steps well, over time (not overnight) you should be getting hundreds and in rare cases thousands of downloads per day just from the app store on a consistent basis. If you are not, reconsider the keywords you chose to target, think about how much search demand there is for those keywords, and how competitive they are to rank for. Try to understand why you are not getting the downloads. Is the problem a lack of demand for the keywords you chose, or a competitive environment that is too difficult?

WHY SEARCH TRAFFIC IS SO IMPORTANT

Most people understand that they need to have their app rank in app store search, but don't necessarily understand all the reasons why it is so important. Let's quickly go over the core reasons why search is so important.

When a person searches for anything, they tell the search engine two things. The first thing is their "intent" (what they want), and the second thing is that they want that thing at exactly that moment. The combination of intent and the right timing makes the person searching a highly converting lead.

Consumers don't just need certain items. They need things at a specific time that works for them. For most kinds of products or apps, the time when consumers need them is a tiny window of time. if you rank in search, you are able to get in front of those consumers at the precise time when they need what you are selling. This makes the consumers who discover your app from search, far more likely to download your app and become engaged users of your app.

Traffic from search is usually more engaged and highly converting than people who randomly discover your app via social media or other means because even if they are interested in your app, they might not have the time for it at the moment you reach them.

Additionally, search is a very long-term source of downloads. If you rank for certain search phrases that bring traffic, you will continue to get downloads from people searching those phrases as long as your app continues to rank. This can last for months or years. Most other marketing strategies do not yield these kinds of long-term and consistent results. If you get publicity or do extra social media promotion, you will have a spike in downloads, but that spike will come back to normal levels within a matter of days, and you won't get similar downloads until the next time you get publicity or extra social media promotion. Doing proper ASO can get you consistent downloads for years to come. Ranking in ASO is actually how I got most of the 1,000,000 downloads for my apps while being free to work on my apps or other projects since after you do your ASO correctly, it is a largely automated marketing strategy where you can be mostly hands off.

Once you do your ASO correctly, you will be able to leave it mostly automated, and you will have time to explore all the other possible marketing strategies that can go hand in hand with your app store search strategies, working to boost one another.

Here is what I do, and what I recommend to others. Start with a focus on ASO (App Store Optimization) by figuring out what keywords you want to, and are able to rank for. Write a proper app store listing title and description that target those keywords. That puts you in a good position to eventually rank for the keywords you need to rank for. Once you do that, you can move on to other marketing strategies. Setting up your app store listing to get the most out of ASO is low hanging fruit that is typically one of the first things I do when I try to promote a mobile app.

HOW TO DO APP STORE KEYWORD RESEARCH

As promised earlier, let's now do a deep dive into keyword research.

Many people are familiar with keyword research because they have been doing SEO (search engine optimization) for the web - mostly optimizing their websites to rank in Google search. ASO (App Store Optimization) keywords research is similar to SEO keyword research, but there are a few key differences.

The first major difference is that a website can have many pages and an app has only one listing page. Each page on a website can attempt to rank for slightly different keywords. Since there is no limit to how many pages your website can have, there is also no limit to how many keywords your website can try to rank for. This is not the case for apps. Apps in app stores (Apple App Store and Google Play Store) get one listing which is usually approximately a thousand words or so. That immediately creates a big limitation on the number of different keywords you can try to get your app to rank for.

You can try to rank for a few main keywords and a few secondary keywords. That is it! You have to choose the keywords to target very wisely.

One additional difference to consider when choosing app store keywords is that people are typing with their thumbs instead of on a full keyboard. The searches people type are usually shorter than what they would type into the Google search box.

This presents a few additional challenges because people searching app stores tend to use shorter search queries that are also narrower in variation and uniqueness. This creates a situation where similar apps have to compete on nearly identical keywords, creating a very competitive search environment.

As an example, let's take a look at the keyword research for my business plan Android app, which ranks #1 on Android for the term "business plan" and the term "business." If this was a website, I could make individual pages that target many specific terms like "business plan for a restaurant" or "business plan for a coffee shop" or "business plan for a home repair business" or any other similar term for every kind of business possible. But on the app stores, people tend to enter a narrower set of shorter search queries which leaves only terms like "business plan" or "business planner" or "plan a business" or "business planning" or "business" that get most of the search demand.

What that means for you as the marketer of your app is that you will be competing for the same few keywords with every other app that is similar to yours. Long-term, this is one of the biggest challenges for your ASO. There are more apps in app stores every day, but not more keywords that they can rank for, and every day the ASO ecosystem is becoming more and more crowded and competitive.

Imagine if you have a photo app. Will you be able to compete with and beat Instagram and other multi-million dollar photo apps? What if you have a puzzle game app, will you be able to compete with thousands of other very similar puzzle apps for the keyword "puzzle app" or similar keywords? Some niches in the app stores are brutally competitive, and it is extremely difficult, if not impossible, to compete with them if your marketing budget isn't in the thousands of dollars per month because for some search terms you literally have to compete in search against billion dollar companies.

Understanding the competitiveness of your business niche, and being realistic about your app's ability to compete within particular business niche keywords should be a part of your business planning even before you sit down to write the very first line of code for your app. If you don't consider these things before you start, you may face a rude awakening months later when you face these facts as you try to get your app to rank in the app stores and find it very difficult to get downloads on a consistent basis.

While some niches may be extremely competitive, other niches may be wide open in which high rankings for your app can be achieved. The problem there may be a lack of search demand. Recall an earlier point about the narrowness of the kinds of searches that are done on mobile app stores. While some search terms get almost all the demand, many other searches see very little demand. Ranking highly for searches that get too little demand is equally ineffective as trying to rank for search terms that are too competitive because in the end, your app gets too few or no downloads.

You can use three keyword research tools to attempt to gauge search demand in mobile app stores. Of course, this data isn't made available by the Apple App Store or GooglePlay, but you can get an approximate idea using the Google Keyword Tool and an app store research tool named SensorTower.

Here is the link to the Google Keyword Tool:

https://adwords.google.com/ko/KeywordPlanner

Here is the link to SensorTower:

https://sensortower.com

Here is the link to Mobile Action:

http://www.mobileaction.co

You can also easily find these tools by Googling for them. The Google Keyword Tool tells you about Google searches, and gives *exact* volumes of demand and keyword ideas on Google. SensorTower and Mobile Action give you *approximate* information of searches and volume of demand in mobile app stores. Sensor Tower and Mobile Action simply don't have exact data, but they have an algorithm to accurately approximate keyword demand in app stores.

With the ability to approximate demand volume and the level of competition, as a part of your business strategy and your marketing strategy, try to find a sweet spot in the keywords that you choose to target where you will be able to compete in app store search for searches that have reasonable demand, but not be too competitive.

As your app grows stronger in app store search, your app will eventually be able to compete for more competitive keywords. So you need a search strategy for the beginning, and a search strategy for once your app matures and will be able to compete for increasingly difficult search keywords.

Of course, don't forget to consider what your users would realistically search for. Think about what they need, explore, or wonder about that might lead them to enter some specific searches. At the end of the day, the keywords you choose must be ones that your potential users would actually search.

Without downloads coming from app store search, growing your app will be extremely difficult. Give keyword research sufficient focus. I'll explain more advanced ASO and keyword research strategies later in this book, and walk you through an example. So if you are new to ASO, don't worry if you don't feel too confident with it just yet. We are not done with it.

Keep in mind, at the end of the book, as one of my free gifts to readers of this book, I offer to give you feedback on your app. If you are struggling with ASO and keyword research, I can give you feedback on the keywords you choose.

SOCIAL AND ENGAGEMENT SIGNALS THAT INFLUENCE SEARCH RESULTS

Once you choose the keywords you want your app to rank for, and add a great title and description to your mobile app listing, you are done with ASO, right? Wrong! This is just the beginning.

What you must do now is work on app store search keyword ranking factors to boost your app in search ranking. On day one, your app will be on the bottom of the search rankings for the keywords you choose, and you must take the right actions to trigger the app stores to rank you higher and higher until over time, you are #1 for the keywords you chose, or close to #1.

Part of your work to get your app to rank higher in app store search will be to trigger "social" and "engagement" signals that app stores monitor.

Engagement signals measure the level of user engagement on your app. Here is a checklist of these signals:

- How many times are they opening your app in total and per day?
- Are they deleting the app from their devices?
- Are they keeping the app?
- How long are the session lengths?
- How many sessions does an average user have?
- Numbers of downloads.
- Acceleration of downloads.

All these metrics show how much users engaged with your app. If users engage with your app more than with the apps you compete with in the app stores, that is a signal that people like your app, and that it is good. If that is the case, this should help your app rank above your competitors in app store search for the keywords that you chose. On the other hand, if your engagement signals are worse than those of your competitors, you should improve your app to make sure that your app is better than your competition on many of those metrics. Of course, you will never know what the engagement metrics are for the apps against which you compete. So just keep improving your engagement all the time. Engaged users will also spend more money with your app. Every small increase in engagement can be a win in many different ways.

Social signals are very similar to engagement signals. Social signals are things like the reviews that your apps gets. The frequency with which your app gets reviews, and quality of the reviews.

Note: the app stores do not officially state that they actively monitor the social and engagement signals of your app. They also do not openly say which of the signals have more or less weight when it comes to influencing search rankings. Almost none of the major search engines (Google, YouTube, Amazon, Yelp, App Stores, etc) make such information public. But they all work on these principles.

As the marketer of your app, your job is to keep working on raising these signals, and monitoring search results for your app to see what starts working to give your app a boost in the search rankings. It is a relatively time consuming process so unless you have a large marketing budget. Do not expect great results immediately. Be patient, and continue to experiment. Over time, triggering the above mentioned signals will help you rank higher and higher in the app stores.

Another thing to keep in mind is that your app ranking has a lot to do with having the right features inside your app, executing those features well, and getting users to actively and consistently engage with them. If that happens, the app stores will interpret that strong user engagement as a very high quality signal, and rank your app higher in search.

Later in this book we will get back to ASO and I will walk you through a practical example of how I achieved success with my own apps. That will help you have concrete takeaways, and reinforce what we went over in this section.

ii. Publicity And Press Coverage For Your App

Publicity and press coverage is when you get featured in different online magazines, blogs, podcasts, YouTube channels, and even TV.

Everyone wants publicity, but less than 0.1% of new apps are covered by major tech or large news publications. Even the apps that do get coverage, either do that through hiring a PR agency (thousands of dollars a month), have great connections (not everyone does so this can be an uneven playing field), or just have incredibly amazing apps. To get press for an app, being very good is not enough. The app and its story have to be amazing because there are already tens of thousands of good apps out there. The bar of quality to get publicity for an app is quite high.

Nevertheless, there are a number of things you can do to generate publicity for your app. Here are 11 strategies to get publicity for your app:

1) GET FREE PRESS WITH HARO

HARO stands for Help A Reporter Out. It is a service where reporters post questions about articles they are writing. If you can become a source (industry term for expert who provides an opinion) or answer any questions that help them with their research, the journalists will credit you as the source in the story and link back to your site. That will help you get immediate publicity for your business as well as get you links from great websites. The links should help your website's Google SEO (search engine optimization), which can help you generate traffic to your website. You can funnel your website traffic to download your app.

Here is the HARO website:

https://www.helpareporter.com/

2) CREATE A PRESS RELEASE AND SEND IT TO HUNDREDS OF PUBLICATIONS USING PRWEB or PRLOG

PRWeb.com is the world's #1 news release service. PRLog.com is another very reputable press release service. They provide lots of free guides and tutorials for how to write press releases and get press coverage. Even their paid products are very affordable. They are certainly much cheaper than hiring a PR firm. They offer templates for you to create a press release and then send it out for you to thousands of publications.

3) USE YOUR CONNECTIONS

A great way to get publicity for your business is to use your connections in your industry. People you know may be connected to journalists, bloggers, or other professionals working at various publications. Your contacts can put you in touch with those journalists.

That can help you get a story about your business published. Unfortunately, few people have those kinds of connections. If you do not have those kinds of connections, it would be a good idea to begin building them as early as possible. Just make sure to ask your friends. You never know who may know someone that can help. It never hurts to ask. You might be surprised by the results.

4) BECOME FRIENDS WITH JOURNALISTS AND BLOGGERS

Since you know that you will eventually need to get press coverage and publicity, research the journalists, podcasters and bloggers who cover your app niche or industry. Try to create relationships with those journalists online since they are probably not all located in your city. It is best to do that before you actually need the press so you can build a little bit of a business relationship with them before you have to ask for a favor. Just about all journalists and serious bloggers use Twitter and other social media so follow them on Twitter or whatever social platforms they prefer, and interact with them. Leave intelligent responses on their articles and posts and stand out by seeming interesting and insightful. When the day comes when you will need press, they will know a little bit about you, and will be more likely to consider writing an article about your business.

5) HIRING A PR AGENCY

Big and midsize businesses typically hire a PR agency or an in-house PR professional. That works great for getting press, but a PR agency typically charges $5,000 to $10,000 per month on retainer which means that you would have to hire them for longer than a single month. So this is not an option for most small businesses. But once your business matures to the point where hiring a PR agency becomes affordable, it is a great option. A good PR agency can get your app or business publicity in top publications and sometimes even television.

6) GET PRESS FROM BLOGS AND SMALLER SITES

Try to reach out to bloggers, YouTubers and podcasters in your niche. If you can, offer them something for free so that they write a review about your app. Many bloggers are much more approachable than large publications. If you have a hard time getting press coverage from the really large sites, try to get publicity from smaller publications in your business niche, or bloggers that blog about business your local area. Each time a blogger covers your business they not only give you publicity and some direct downloads, but they also link to your website which helps your Google SEO.

You can sometimes pay bloggers, journalists or podcasters to give you coverage. It is usually very affordable if you target relatively small publications. If your app generates revenue, you may be able to earn back in app revenue what you paid to get the publicity.

7) ANOTHER STRATEGY TO GET ON PODCASTS AND RADIO SHOWS

Appearing on podcasts and radio shows is another great way to get press coverage. Just like HARO is for journalists, a site called RadioGuestList.com is for podcasters and radio show hosts to look for guests. The RadioGuestList.com website sends a daily email which lists podcasts and radio shows that are looking for particular types of guests. If your business background matches what the podcasters are looking for, you can email them and try to book an appearance. Many radio shows and podcasts also have websites on which they typically post show notes. If you get yourself an appearance on a podcast, that show will also likely post an article about the episode with a link to your website, which can help to boost your SEO as a secondary benefit.

This particular strategy has been so effective for me personally that I made a full course about it online. Within one year I booked myself 50+ appearances on different radio shows and podcasts for free and without hiring a publicist. You can get this course for free because as one of the free gifts I offer you at the end of this book is access to one of my online courses. You can choose to get this course for free. Here is the link to the course so you can browse the full list of lectures:

https://www.udemy.com/how-i-got-50-podcast-appearances-using-radioguestlist/

8) PRESS COVERAGE BY GUEST BLOGGING

In case bloggers do not want to cover your app, you have another way to get publicity from their blogs. You can ask them if they would accept a guest post. A guest post is an article that you would write that would appear on their blog. In a guest article, you can link to your website. That may help your SEO, and potentially get you app downloads from people that come to your website from those blogs.

Few suggestions on guest blogging: make sure that the publication in which you are trying to place a guest article is in the same business niche as your app or business. The content must be relevant to their readers and your audience.

Additionally, don't over-do it with guest-blogging. A few times a month on different publications should be max. And always try to guest-blog in high quality blogs. Don't waste time placing articles in tiny blogs or low quality websites because they will not get you traffic, and the links from them will not be valuable for SEO.

9) STAND OUT AND BE EXTRAORDINARY

One thing to always keep in mind and think about is how your app, business, and overall message can be or be more unique, interesting and unusual. When your business stands out, you are more likely to naturally get noticed and get press coverage.

10) PUBLICITY BY JOINING A PRESS CLUB

Every major city has a press club. Search Google for a press club near where you live, go to the press club in your city and meet various journalists there. You can tell them that you are a source. A source is the term journalists use when they describe an expert or someone who contributes to their stories. If the journalists in your local press club use you as a source, they will credit you in the articles they write by mentioning your name and your business. That will not only give you publicity, but will also help your SEO efforts because they might also link to your website or blog in their story.

11) PUBLICITY STUNTS

Another strategy you can use to get free publicity is to use publicity stunts. You can do something extraordinary and unusual that grabs attention. This is very similar to Seth

Godin's concept of the purple cow where you should do something unique and extraordinary to stand out and get attention.

iii. Social Sharing From Social Networks

Most first-time entrepreneurs understand that they need to be marketing via social networks, but do not quite manage to do it effectively. There are two general keys to effective social media marketing:

1) Leveraging social media influencers in your niche.
2) Turning yourself into a thought leader with an audience.

To think about social media marketing more broadly, consider that blogging and guest blogging is also part of social media. Additionally, think about sites like Quora, Reddit, HackerNews and other sites in your niche where content is shared.

If you are in your business for the long-term, when you think about social media, think about how you can position your business and yourself as thought leaders and an authority in your niche or industry. That can happen by either getting publicity or by establishing your own large following on large social media sites.

As examples of thinking more broadly about social media, you can start your own podcast or a YouTube channel. In fact, I started my own YouTube channel to promote my apps. The beauty of creating your own large following on social media is that if you manage to grow your social presence to be big enough, it will drive traffic and downloads to your apps daily. It will be almost like getting publicity for your apps every day!

iv. Social Sharing And User Invites Generated From Inside Your App

As people use your app, there will be some things that they can do together with friends or people they know. That can be playing a game together, sharing content, or as the case in my business plan apps, users can collaborate with business partners and write their business plan, marketing plan or a fundraising plan together.

When you plan features for your app, try to come up with features people would want to use with people they know. Enable and encourage your users to invite their friends to use those features on the app. This will get you a boost in downloads from your users bringing their friends into the app.

As an example, think about a game like Words With Friends. One person who loves playing that game can invite hundreds of their friends over time. Then those friends can invite their friends. Over time, a single download can generate hundreds or thousands of other downloads through people inviting their friends from within the app. This is the ideal scenario for mobile apps. Your job is to think about how to position your app to make such viral sharing possible. What feature can you build that will make it natural for your app users to invite their friends to download your app?

v. Website Promotion

Your website should also be generating downloads for your app. You can drive people to your website via Google SEO, social media, answering questions on Quora.com, posting content on Reddit, Twitter or Facebook, participating in other sites and communities within your niche, or via many other strategies that get you web traffic to your website.

Many app entrepreneurs treat their website as a big landing page that has a big download button. But there is so much more you can do with your website. If you start creating content which also promotes your app, you can promote that content with Google SEO and all over the web. This can generate downloads with each piece of content you produce.

You should optimize each page of your website for app downloads by adding big buttons that are calls to actions for people to download your app.

There are two options for where your blog can exist. You can maintain a blog on your site, or you can have an entirely different site for your blog like I do with GlowingStart.com and Problemio.com. The advantage of having a different site is that you can try different promotional strategies on the two different sites, and compare results. Plus, the topic of your blog will only need to loosely match the topic of your app if they are promoted via different sites. That will give you more creative freedom.

The bad thing about having two websites is that you have to do everything twice, it is a management headache, and your branding gets diluted.

vi. Cross Promote And Partner With Other Apps And Websites

This is one of my least favorite approaches, but it can work. Simply find partners that are willing to cross-promote with you. You can drive traffic to them, and they can drive traffic to your app. These can be mobile apps, websites, podcasts or anything else. The key is that you should get as much exposure as you are giving to your partners. Remember, whenever you are sending one of your users out of your app, and toward your partner's app, you are killing engagement of your app, which is damaging your app store optimization (ASO). For that reason, this is one of my less favorite strategies. Think twice about whether you want to do this.

A better approach is for you to make multiple apps which you can cross-promote to your users. For example, I built my apps as a 4-app course of 4 related apps. As many as 15% of the people who download one of my apps go on to get at least one more app. This way you immediately maximize your downloads without improving any of your outside marketing.

Having multiple apps also allows you to target more different keywords in the app stores by having two app store listings.

vii. Pay For Advertising

You can certainly get downloads by paying for ads on mobile apps and websites. This can be a viable way to promote your app, but only once you have established your revenue model and have an understanding of how much money you earn on average per download.

If you are able to generate enough revenue per customer to cover the costs of buying new customers via ads, then paying for advertising is very viable. But for most apps that are starting out, this isn't the best option because those new apps typically don't generate enough revenue to cover the costs of the ads. This is only a great option once your app generates more money per user than it costs to acquire them.

viii. Email Marketing And Growing Your Email List

As part of maintaining a blog or via features in your app, you can collect the email addresses of your users. That will enable you to reach out to them whenever anything noteworthy happens and you need to make an announcement about it.

For example, if you have an app update, a new blog post, a new YouTube video, create a brand new app, or anything else, you can send an update to your email list and drive them to your new offering. In addition, you can also make money from your email list by notifying them of things you are selling, or other kinds of offers.

iv. Speak At Events

Speaking at events can be beneficial to your business in a number of ways. The obvious benefit is that audience members may download your app. But that is only a minor benefit compared to the full potential of speaking engagements.

If there are reporters or bloggers in the audience, they may include your app in their story. That will get you even more downloads, links for SEO from their articles, and possibly give you a social media boost in followers. Additionally, the conversation of the evening will be focused on you and your app, which can result in plenty of feedback about your product, further introductions and beginnings of interesting business relationships.

v. Create A YouTube Channel

Having a YouTube channel makes sense for a number of reasons, especially if you are planning to stick with your app as your main business for a long time.

First, YouTube is the second largest search engine on the English-language web. Many potential new users can discover your app on YouTube, and you can drive many of those potential users from YouTube to your app.

The next best thing about YouTube is that its videos can be re-purposed in many ways. If you want, you can place your YouTube channel inside your app. Having videos people watch inside your app will help you increase time spent on the app which is an app store ranking factor. You can also embed the same videos on your website, which will increase the time people spend on your website, and possibly the ranking of some of your web pages that have your YouTube videos. Additionally, your YouTube videos may show up in actual Google search results. And if all that wasn't enough, YouTube videos are more likely to be shared because people like videos, and it is often easier to consume video content rather than reading text. Plus, if all else fails, you can make money with YouTube videos by placing AdSense ads on them.

YouTube is possibly the most versatile social media platform, and your videos can be used for many purposes.

Another YouTube tactic you can try if your app is a game is to put highlight or advanced tutorial videos of how to play your game on YouTube. That can funnel some of your app users to your YouTube channel, which will help you grow your YouTube channel. Once you grow your YouTube channel, you will be able to drive people from YouTube to download your app.

vi. Keep Your App Freemium

Have you noticed that most apps in the app stores are free? That isn't because developers don't want to make money. It is because the app stores make it much easier to promote free apps. When people browse apps, they will almost always choose the free available apps over the paid apps. It is also much more difficult to get a paid app to rank in app store search, and many consumers never discover most paid apps.

The common approach is to make your app free to download in order to make it competitive in terms of marketing on the app stores. Once your app is competitive in app store search and is getting downloads, you will have to figure out how to make money from your users as they are using your app. Later in this book I'll cover the effective ways to generate revenue from your users.

vii. Google Search (SEO) For Your App Store URL

While you cannot track how many people click or see your app's URL in Google search (not your website URL, but the web URL of your actual app which is the web iTunes or GooglePlay listing of your app), it does show up in various Google searches and you can do SEO for that URL in order to get it ranking well. In fact, your app's iTunes or GooglePlay URL will often outrank your website because it is on a very authoritative domain which is iTunes or GooglePlay.

To get your Apple App Store or Google Play URL to rank higher in Google search, try to point some links to it. The simplest thing to do is to write a number of blog posts for your blog, and in each of those blog posts to link to the app URL. As you get publicity and blog mentions, whenever possible, get them to link to your website URL and your app URL to help both rank higher in Google search.

If you are slightly confused by what I mean by web URL of your app, here is the example of my app's web URL:

https://play.google.com/store/apps/details?id=com.problemio&hl=en

viii. Establish A Presence On Any Large Platform

Just like establishing a YouTube channel may be a great way for you to generate downloads for your app, any other large platform may be equally as good as long as you are able to establish a large presence on it. Think about what platform may have the most natural audience for the kind of app you have. It might be some big community in your niche or a big industry website or a large following or Pinterest or Facebook or Twitter, or anything similar.

xiv. Become A Great Marketer

Whether you plan to work on the current app for years, or build other products, whenever you build something in the future, to ensure its success, you will need to promote it like an absolute pro. The better you become at marketing and sales, the more products you will sell. For that reason, invest some time and effort into becoming a great marketer by learning and practicing it. It will help you for the duration for your career as an entrepreneur, far beyond the current app project.

xv. If Your App Is Android, Release It On Multiple App Stores

There are many Android app stores that are not Google Play. There is one for China, a number of other countries, and many smaller app stores ran by different technology companies like Samsung. The Amazon Kindle also has an app store where you can add your Android apps. The NOOK platform also has its own app store.

One of the biggest factors in deciding whether to post your app in other app stores is how your app makes money. If your app makes money with ads, you can display those ads to people all over the world, and it makes sense to post your app in app stores of different countries. On the other hand, if your app makes money by selling in-app purchases, you will have to implement the in-app billing differently for every app store on which you plan to post your app. If you post a slightly different version of your app on many different app stores, it will cause you to have a management nightmare because you will have to make a single app update many times, one for each app store version. This can slow down your development process, and hinder product improvement over time since you will have to allocate resources to maintenance more than development.

xvi. Participate In Forums

Posting on forums and telling people about your app is another effective marketing strategy. Just don't confuse participation on forums with blatant spam. What you don't want to do is join many forums, and immediately put links to your app on all of them. People don't like that, and there is a large chance that your posts will be deleted if you are perceived as a spammer.

Instead of aggressively posting links to your app on forums, start by becoming a valuable contributor on forums where your potential users hang out. Gain trust on forums first, and once you are a regular member there, you can sometimes point people to your app.

xvii. How To Ensure That You Get Good App Reviews

Mobile apps live and die by user reviews. If an app has good reviews, new potential users see that, and download the app too. If the app has bad reviews, new potential app users see that, and because the reviews are bad, avoid the app. Thus, having bad reviews decreases the number of downloads your app will have. And having fewer downloads will make you weaker in the app stores in terms of ranking because your competitors will have higher accelerations of downloads, and will beat you on this metric. This will help them jump over you in the app store rankings over time.

Now that you understand just how necessary it is for your app to have amazing reviews, let's discuss how you can ensure that your app has great reviews.

Apps that are successful at getting good reviews typically have one thing in common. They ask users to leave reviews. There is a little bit of an art to asking users for reviews at just the right time and place in your app to get many users to give you nice reviews.

You must ask users to review your app when they are very happy with your app. If you ask users to review your app at random times, you will get good and bad reviews. But you only want your app to have good reviews. You must understand the experience of your app users deeply enough to know exactly when your users are happiest with your app. For most apps, this is when your users have been using your app for a while. They can't possibly hate your app if they are regular users of the app, right? Think of areas of your app where very engaged users end up, or the features your super users use, and ask for reviews there.

Another instance when it might be a good idea to ask users for a review of your app is when they have an "aha" moment, or get what they need out of your app. For example, in my business plan app, I ask users for a review when they finish reading the tutorial for how to write a business plan. The assumption is that if they get to the end of the tutorial, they must have felt like they were getting something out of it. Otherwise they would have quit in the middle. This assumption proved correct, and over time, I have gotten hundreds of 5-star reviews from that exact spot of my app.

You can also get your friends, business partners, family members, and people you meet to use your app and give it a 5-star rating if they like the app.

Additionally, many app developers buy 5-star reviews from app marketing agencies. I don't recommend doing this, but many app developers do use this tactic, and it can work very well. Of course, you should be aware of the risk. If the app stores begin penalizing people for having many fake reviews, who knows how harsh the penalties may end up being. This is obviously very risky. I prefer to make my apps so good that they naturally get good reviews from happy users.

xviii. How To Avoid Bad Reviews In The App Stores

Every app entrepreneur wants to avoid bad reviews of their apps in the app stores. A user may give your app a bad review, leave, and forget about your app. For them it is not a big deal. But this bad review will damage your app's reputation for the lifetime of your app, doing its part to bring down your review average, and discourage new people from downloading your app. Any bad review you can avoid will translate directly to you getting more downloads and generating more revenue from the maximum potential downloads.

Of course, it isn't possible to avoid 100% of the bad reviews that may come your way, but you can avoid a very large portion of the potential 1-star and 2-star reviews that you may potentially get by following a few simple strategies.

The first and most obvious strategy is simply to make sure that you create a good app that is well tested and free of any major bugs. I realize that this goes without saying, but mobile app users tend to be especially finicky. They expect even free apps to function perfectly. If your app isn't useful, or if it crashes too frequently for their liking, they may give your app a bad review. Even if your app is free, put a lot of focus on ensuring that your app satisfies your customers above and beyond their expectations.

When I coach new app entrepreneurs, I tell them to make sure that their app is as good as it can possibly be, without crashes and with good usability. But that is easier said than done. Many apps that are made by new app entrepreneurs I coach aren't as good as users need because this is the first time these entrepreneurs have built a product. They often make many beginner mistakes.

Product quality is one of your most important marketing tools for almost any business. If the product is good, people will engage with it more, recommend your app to friends, and leave good reviews. But if they don't like the app, they won't recommend it to friends, and leave bad reviews, which will discourage future potential users from downloading your app.

Since you will never fix all the bugs in your app and some defects are bound to creep in, another tactic to avoid bad reviews is to add your tech support email address at the bottom of your app description in your Google Play and Apple App Store listing. Some of the people who experience technical problems with your app will see your tech support email as they try to leave a bad review, and email you for help instead of giving your app a bad review and disappearing.

Another strategy to make sure that you minimize your bad reviews is to ask for reviews in parts of your app that are less prone to bugs and where you are certain that your users will be happy with your app. In the case of my business plan app, I added content articles which are tutorials for how to write a business plan. These articles are helpful, and are not prone to bugs since the tutorials are just text-based content and not a tool where bugs can creep in. Conversely, I deliberately did not place any calls to action for users to add reviews in the actual business planning part of the app because if the users disliked the usability, felt the app lacked some features, or the server was down, they may leave bad reviews. I just didn't risk asking for reviews there.

Another strategy to minimize bad reviews may be one that you find surprising. Whenever you or your business interact with anyone online, make sure you come across as positive and non-controversial. You never know who you might anger online if you argue topics like religion, politics or anything else that can be controversial. If you get into an argument with the wrong person, and they decide to slander your work, your app will pay the price.

Another strategy to minimize bad reviews is to not put too many ads on your app, and not to up-sell anything too aggressively. If your users take offense at your ads or overly ambitious sales of anything else, they may leave a bad review. This point goes hand in hand with making sure that your app or any in-app purchases are not overpriced. If users think something is too expensive and doesn't offer enough value, they may leave a bad review.

xix. How To Boost Social And Engagement Signals Of Your App

The app stores monitor how users use your app, and use it to gauge the app's quality and relevance to people's needs. The data they gather is broken down into engagement and social signals.

Every app developer wants to increase the time users spend with their app. Engagement a great proof that the app is a good product, it helps with app store rankings, and the more time people spend with an app, the more likely those users will warm up to the idea that it is OK to spend money buying something on the app. I'll share some of the things I did to help my apps have better social and engagement signals than my competitors.

The first thing I did that helped my apps get great reviews in the app stores and increase the social engagement is that I enabled chat in the business planning app. I added a feature to the apps where users who were planning a business could ask me questions, and I would answer their questions right on the app. Sometimes we would get into long conversations and many questions would follow people's initial questions. Each comment, of course, meant that they would have to open the app (engagement signal) and write a comment (extra time spent on the app which is another engagement signal). But the best part of this was that my users got truly amazing personalized help planning their business. Because of that, people were thrilled about the level of help they were getting from the app, and were leaving great reviews about the app (social signal). This single feature helped the app boost many of its social and engagement signals. Whatever your app might be, if you can add some sort of an interaction either between you and your users, or allow your users to interact with other users (this is even better because you don't have to be involved), it can help you significantly boost your engagement signals.

Another thing you must include in your app are sensitively conceptualized push notifications. Push notifications can feel very intrusive to users, so you must be very careful not to overwhelm or spook your users with push notifications. If you are too aggressive with your push notifications, it will backfire, and people will delete your app.

Your push notifications must always be welcomed by your users. To accomplish that, you should only alert them of things they opted into, and that they truly look forward to. In the case of my apps, I only use push notifications to alert people when they are getting help with their business if they requested help or asked a question. Every time you get a user back on your app, it is a little win for your engagement metrics and your app store rankings. But be careful not to overdo it with the push notifications. If they annoy your users, those users will delete your app or add bad reviews, and you will lose any ASO benefits or any chance to make money from that user. So be very careful with how you use push notifications.

Think of features you can add to your app where users engage with one another. It can be something like a turn-based game where every time it is a user's turn, they get a notification. The game itself can engage users by giving them extra free points every day with the user having to open the app to accept the extra points. That gets the user to open the app daily, and gets them to re-engage with the app. Messaging features have this advantage "built in" because every time a user gets a message, they get an alert that they need to open the app to see the message.

You can notify users in two ways. You can use email or push notifications. Push notifications are more effective to get users into your app, but you can't use push notifications in every case. When you can't use push notifications, fall back on plan B, which is notifying them via email. On Android, users must opt into push notifications when they download the app, and on iOS, users have a choice of whether they want to receive push notifications from your app or not. If some of the iOS users opt out of getting push notifications, have a way to collect people's email to ensure that you can still message them and remind them to get back into the app.

There is an important nuance when it comes to the Android platform. To enable push notifications, your app must require extra permissions which are asked as the user is deciding whether they should download the app. Some of these permissions may feel intrusive to your users, and they may decide not to download the app due to the extra requirements. When you add push notifications, make sure that they bring a significant enough amount of benefit because to enable them, the extra app requirements will cause a slight decrease in your overall download numbers. Just keep that nuance in mind when you are making the decision of whether to enable push notifications on Android. Although push notifications are effective at getting people back into your app, there may be some cases when it makes sense to decide to notify users via email instead.

Inviting friends is another strategy that can help you boost your engagement signals. Adding social features to app is great because it helps you get more downloads. A great side effect of inviting friends and getting more downloads is that your download acceleration goes up. App stores track that, and you begin to outperform your competition on this metric, which helps you rank better at the same time.

Another interesting trick I like to use in my apps is embedding video that can be watched inside my apps. If an average session length for an average app is about a minute, and you get people to watch a video that is five minutes, you just increased your user session length average, and thus beat your competition on that metric.

In my own apps I took this to an extreme. Since my apps cover business ideas, business planning, marketing and fundraising, I started a YouTube channel covering these topics in depth. Over time I have been released over 800 videos on my YouTube channel, all of which can be viewed inside my apps. If embedding videos makes sense for your app, try it to help you boost your average session lengths.

Another thing you can do is have new content be available on the app on a regular basis. That will give users a reason to regularly come back to the app for the new content.

Since I release videos regularly and it regularly has new videos to watch, that gives people a reason to keep opening the app to watch new videos. This helped me raise a few engagement metrics by getting users to re-engage with the app and deleting the app much less frequently.

If you do something with video, your app users can subscribe to your YouTube channel, which will help your brand grow on YouTube, and make your videos more prominent there. That will help you grow an independent large presence on YouTube that you can then use to funnel people to download your app, or sell any other product you have.

A fascinating thing about YouTube is that the audio from the videos can be taken and made into a podcast. This way you can increase engagement on your app by embedding your podcasts just like you embed your videos, and also grow a podcast audience on iTunes that you can then funnel back to your apps, just like you did with YouTube.

If you are curious, check out my YouTube channel here:

http://www.youtube.com/user/Okudjavavich

NOTE: One important nuance to keep in mind is that you must embed the podcasts or YouTube videos right into your apps. Don't just link to that content and allow users to leave your app, but make sure that people are watching embedded content while they are still on your app. If people leave your app to consume that content, this will not help your engagement metrics. In fact, it will hurt your engagement metrics because people will be leaving your app. So YouTube or podcasts have to be embedded as a part of your app. To see how I did this with my apps, you can check out any of my apps on http://www.problemio.com and use my example to guide you on how to embed this kind of content in your app.

Another tactic you can use to improve the engagement metrics of your app is by adding written educational materials or tutorials. This doesn't apply to games, but if your app is a utility or a business app, or an app that serves some other practical purpose, most likely your app users can benefit from a little bit of education about the subject matter. Having your app users consume written content will also increase the time your users spend on the app just like with watching videos or listening to audio content. Written content is also great because you can code that into your app, and your users won't need an Internet connection to consume that content like they would if they wanted to watch a YouTube video or listen to a podcast episode. Plus, if you have written, audio, and video content on your app, you can be sure that you are satisfying many different kinds of people who like to consume different types of content. No one will be able to complain about your app not having the kind of content that they like.

Let's recap the strategies for how to continuously get new content.

- You can engage with users personally (it can be powerful, but I rarely see this on apps outside of my own).
- Users can engage with each other.
- You can create original content that your users can consume. The more this can be done without your direct involvement the better because it will save you time to work on other things related to your app.

I also want to go over the advantages of having a marketing budget. If you are able to pay to enhance some of your marketing efforts, your life as an app marketer can be much easier. For example, remember how increasing the acceleration of downloads can help you boost your engagement signals and give you an advantage over your competition in app store rankings? What if you were able to afford either hiring a PR agency or paying $1-2 per download (that is close to the going rate) for thousands of downloads? If you are an individual developer or a part of a small team, you probably can't afford that yet. But if you find yourself in a competitive niche, some of your competitors will have deep pockets. They will be able to superficially boost their engagement signals by paying for downloads this way.

To do well in competitive app niches, you may have to raise money to have a chance. And raising money isn't for everyone. Most people don't succeed at raising money. So make sure you take all this into account when planning your app.

xx. 3 Ways To Launch Your App

Launching your app is one of the most exciting moments in your journey as an app entrepreneur. There are only a few other moments in that journey when things will get more exciting. Those moments will occur when you figure out how to get the hyper growth you are after, when your revenue and profitability will reach and surpass your goals, and if you ever sell your app for a lot of money. It is fun to dream, isn't it? For now, let's come back down to earth, and make sure that you get your app's launch right.

There are 3 ways to launch your app. You can do a quiet launch (this is what I did), a launch with lots of publicity and attention, and an extra loud launch on stage at a conference. Let's cover all three in detail.

SOFT LAUNCH FOR YOUR APP

A quiet launch (sometimes called a soft launch) is a launch in which you simply make your app available in the app stores of your choice and only mention it to your friends, family and some professional contacts. A small number of people will also find your app directly in the app store.

The point of a quiet launch is to have only a small set of users who will be your initial beta user group. This is crucial because you will need to observe how they use your app, and use analytics to get an understanding of how your initial users are using your app, what they like about it, what they don't like, and at which points of the app users quit using your app. Doing this will give you insight about where your app is strong and where it is potentially weak or buggy. You can use that insight to improve your app until it is ready for a bigger audience. After you improve the app, you can quietly release an update of your app to do more testing and monitoring of your users to determine next areas of the app that need to be improved.

This iterative process of improving your app should never stop, and you should always devote resources to improve your app in the manner I just described. Product quality and customer satisfaction are some of your biggest potential advantages when it comes to creating a great business.

At one point you will decide that the app has been improved enough to be promoted to a wider audience. This is when you can start thinking about doing a big launch with publicity.

HOW TO DO A BIG APP LAUNCH WITH PUBLICITY

Throughout the soft launch of your app you should be creating relationships with journalists, bloggers and podcasters who cover apps or your specific industry, mobile app review site editors, other app entrepreneurs, and YouTubers who focus on apps.

If you are successful at getting them excited about your app, when you are ready to do a big launch of your app, you can reach out to them and tell them that on a given date you will launch your app. A few of those people will help you, and suddenly you have your launch date set, on which you can get a boost from extra publicity. If you are able to allocate a marketing budget for your app, it will go a long way to convincing people to help promote your app during your launch, significantly boosting your launch efforts.

Having a big launch is great, but for long-term success you need to have further strategy. As you already know, most apps rely on app store search for many of their downloads as their main long-term marketing strategy. You need to incorporate that into your launch strategy by positioning your app to get a search rankings boost from your big launch. To do that, you must come up with a good app title and description that target relevant search terms via which users will find your app when searching. You can use the download burst from your launch to get your app to rank for those keywords because one of the metrics app stores use to generate search results is the acceleration of downloads for apps.

Let me explain this a little bit further. Recall that app stores track metrics like app download accelerations, positive reviews, and time spent on the app. Well, this is your chance to boost your app download acceleration by orders of magnitude via the publicity during your big launch. You may be wondering how the publicity will help to increase the time people spend on your app. Publicity won't help that on its own. This is where your soft launch and your beta testing should play its part. User engagement is one of the core aspects of your app that you must work to improve during your beta testing.

Ideally, by the time you do a big launch of your app, user engagement should be in a good place. If your publicity efforts result in hundreds or thousands of extra downloads, it should (no guarantees) help you immediately achieve relatively good rankings in app store searches. Of course, if your app is in a very competitive niche, it won't be that simple. Nevertheless, this is the strategy to give your app a nice rise in app store search rankings from the beginning, by having a great launch.

The kind of a launch I just described should go a long way to getting many downloads and helping your app rank well in app store search. Nevertheless, it doesn't always work out perfectly. App entrepreneurs sometimes come to me after doing a launch like this, and ask me why their app isn't ranking well. Most of the time the problem is easy to spot. The most common issue is that there is a big disconnect between what the app developers or app entrepreneurs think is a point where their app is good, and the opinion of their users. When I download the apps of those entrepreneurs, I can usually spot places within the app where users would get confused, and stop using the app. Often, since their app is new, the app is so poor that is it barely even usable.

User confusion is one of the biggest reasons users quit using apps. It might seem harmless to us app entrepreneurs if the app usability is not quite perfect, but when your app users are confused they subconsciously feel stupid and bad about themselves. That causes them not to go back to the product that made them feel this way. You might think that they should just spend a little bit more time trying to figure out how to use the app, but many of your users simply won't do that. They will quietly quit without telling you why. Or even worse, some of those users will tell you about it by leaving a bad review about your app. You certainly don't want to find out about why they quit your app this way.

During your soft beta launch and moving forward, you must make sure that your app usability does not confuse people.

This is easier said than done. To be completely honest with you, I still have spots in my apps where users get confused. Despite me knowing where my users get confused, to fix or improve those areas isn't always easy. It can require rewriting a large part of the usability without guarantee that the next version of the software will be any more user friendly, and at the risk of introducing new bugs. Nevertheless, try to minimize confusion points within your app as much as possible.

When your app's usability is smooth, there is minimal confusion, and you are able to accelerate downloads, you should be in good shape in terms of app store search rankings as long as your app is in a niche that is not ultra competitive.

BIGGER APP LAUNCH: PRESENT YOUR APP AT A CONFERENCE

There is one strategy to take your big "launch with a boom" and amplify that boom further. That strategy is to launch your app on stage at a start-up conference. This isn't an option for most app entrepreneurs, but if your app is unique and intriguing enough, consider applying to launch at the Launch conference hosted by Jason Calacanis, the TechCrunch Disrupt conference, any of the other major technology and app conferences, or a conference focusing on the particular industry of your app. Presenting on stage at a conference will not only expose your app to the immediate audience of the event, it will also inform the reporters in the audience about your app, increasing the chances of getting press coverage in the publications for which they write.

xxi. Translate Your App's Title And Description To Other Languages

By now you are aware of ASO and app store keywords. But did you realize that people might be using the same keywords but in different languages?

For example, if the keyword you are targeting is "hello" then only the people who search the app stores for "hello" will find your app. But what about people searching in different languages? What if they search for "hola" which means hello in Spanish or "bon jour" which means hello in French or if they search for "hello" in any other language.

As soon as you translate your app's title and description, your app will begin appearing in app store searches that are searched in other languages. People sometimes report over 100% immediate increase in downloads after translating their app titles and descriptions into other popular languages. The best part is that you can get immediate free translations using GoogleTranslate.

Chapter 3: How To Monetize Your Mobile App

Monetizing your mobile app is probably the most difficult part of creating a successful mobile app business. The focus on monetization should begin during the business planning phase, before you even write your first line of code. There should be natural, apparent, and proven ways for how to make money from the mobile app you are planning to create. Having said that, the best time to put most of your focus on monetization is when your app begins to consistently get many downloads. If your app has many downloads, that is when you are going to be able to make any kind of significant revenue from it. There isn't much sense in trying to make money with the app when it first launches because if you don't have many users, the app won't make much money anyway.

i. Should A Mobile App Be Free Or Paid?

The trend for mobile apps is to be free. More and more apps are free, and the app store ecosystems are geared for apps that are free to get more exposure, and ultimately win. Just imagine, if you are a user who is searching the app store and scrolling through the apps in the search results, are you going to try the free ones first, or the paid ones? Nearly all consumers first try the free option, and rarely get to the paid option. When you have a free and a paid product next to each other, most consumers gravitate towards the free offering first.

Most apps today (and increasingly so in the future) will be free and operating under the freemium business model.

Most of the highest earning apps are actually free. They earn money by generating a significant amount of downloads and finding ways to make money from their users as they are using the app.

ii. How Much To Charge For An App If You Do Make It Paid

Let's explore some of the thought process behind finding the ideal price for your mobile app.

The problem with making your apps free is obvious. As the developer, you probably prefer to make money from from your work. You didn't slave over your app just to give it away. And there is no guarantees that your free app will be lucrative because it is entirely up to you to experiment with different monetization strategies inside your free app, and there is no guarantee that you will ever find a great monetization strategy that is as effective as you need.

Despite the app stores ecosystem pressuring developers to make their apps free, there are many different situations when you should make your app a paid app. Let's consider them in no particular order.

One case when you can make your app paid is if you are able to get distribution from another platform. For example, if you have a large YouTube channel, you can funnel people from YouTube to buy your app. If you think about it, the whole point behind making your app free is to put it in a position to get downloads from app stores. If you can get mass distribution from any other platform, and are able to funnel people from there to your app, you don't have to rely on app stores as much, and can make your app paid.

Another instance when you can make your app paid is when you don't have the desire or resources to experiment with in-app monetization strategies. It can take a significant amount of your resources (time and money) to find an effective in-app monetization strategy, and there is no guarantee that you will ever find one. For that reason, you can just make your app paid.

Another case when you can make your app paid is if it is something very unique. For example, there is a famous app that lawyers use to study for legal exams. That app is paid, and it is expensive because it is a one of a kind app.

Another instance when an app can be paid is when you sell it to your existing customers as a product in your product line.

Another instance when an app can be paid is when the niche in which you are competing in is not competitive, and the app does not need to compete in the app store against similar apps that are free.

HOW MUCH TO CHARGE FOR YOUR PAID APPS

Apps live and die by their reviews. What that means for your app price is that your users must feel that they are getting a good deal when they buy your app. So you are forced to maintain low prices. The lowest price range is the $0.99 to $1.99 with 2.99 already considered mid-range for app pricing. These prices are very problematic because it is difficult to get people to open up their wallets, and when they do, you are only able to extract a few dollars. If a customer is ready to spend more money with your business, you have no way of making that happen.

For an app to be $4.99 or above, it should be quite special, or consumers will revolt by posting reviews that are not very good.

iii. Making Money With Ads

No matter how you spin it, your users will dislike (or sometimes hate) the ads that you put in front of them. And if users hate your ads, the reviews they will give your app will be worse than if your app did not have those ads. That is just a reality of monetizing your apps.

If you decide to use ads, you can use banner ads, interstitial ads, ads to download other apps, or in-content ads that don't look like ads, but look like your own content. The latter two are my favorite strategies, and ones that I recommend if you decide to publish ads on your app.

The effectiveness of an ad's monetization is measured in CPM which is the money you earn per 1,000 impressions of that ad. When you hear different vendors talk about the effectiveness of their ads, they will most often refer to the CPM of the ad.

Let's explore the different ads you can use to monetize your app. While I discuss different ad types, I specifically do not mention any companies or vendors.

Banner Display Ads

The simplest and most familiar type of mobile ad is the rectangular banner ad that appears on the screen while users use your app. Web and mobile users are well trained to avoid that ad, and develop a certain ad blindness to this kind of an advertisement. Almost no one clicks on such ads, and when they do, a large part of those clicks are mistake clicks where the user meant to tap on something else, but hit the ad instead due to the small screen sizes of mobile phones. These ads are usually disliked by users, earn low CPMs, and result in decreased engagement and worse reviews for your app.

Interstitial Ads

Interstitial ads are ads that pop up when you are in the middle of doing something like going from one screen to the next. They monetize a little better because they are more "in your face" to your users, but that is also why most of the time, users hate those ads with a passion. A very large part of mobile users scramble to find how to close that ad as soon as it appears. I personally do not recommend using such ads as I had very poor results when I experimented with putting them on my apps. Nevertheless, many app developers use them because they can be more effective at generating revenue than basic display ads.

Ads To Download Other Apps

One typically effective way to use ads on your app is to advertise other apps on a pay per install basis. A typical install, depending on the kind of app it is, can earn payouts of a few pennies to a few dollars. If you can promote the more lucrative apps, this can be an effective monetization strategy. To find such offers, just search for "Pay Per Install" or CPA mobile advertising firms.

In-Content Ads

Another kind of ad that I like doesn't look like an ad at all. It looks just like the rest of your content. You can promote anything you want with that kind of an ad. It can be in-app purchases, your other products, or affiliate offers. The key is to make the text explaining the offer, and the button that links to the offer to look like it naturally belongs in your app, and is a part of your content.

For example, one of the products I offer on my apps is for people to create a website. Most people starting a business need a website. So in my apps, I made a tutorial for how to set up your website for free and without needing technical help. My users like this article because it is truly helpful. A part of that tutorial focuses on web hosting, which is something all website owners need. There, I suggest a particular hosting company, and link to it. It is helpful to my app users, and it is a natural thing to promote in my apps. My users appreciate it, and I get to make money. That is the ideal way to create an in-content ad.

One note of caution is that you must always disclose to people that this is an ad or a partner company of yours, and that you will gain something from them engaging with whatever you are promoting. You can make that disclaimer subtle somewhere near the ad.

Mobile App Mediation To Earn Hundreds Of Percent More From Your Ads

If you use an off the shelf ad company like AdMob or something similar, you should use something called ad mediation. This is where you use a 3rd party software which aggregates about 20 services like AdMob and gets you the best ad to display to users each time. This can result in doubling or tripling your revenue from your display ads.

I don't promote individual vendors in this book, so simply Google for "smart ad mediation for mobile ads" and you will find many companies that provide this service. It is easy to set up and you can begin making more money from ads almost immediately as soon as your app update with ad mediation hits the app stores.

iv. Making Money With Affiliate Products

You can sell products made by other companies, and earn a commission. That is called affiliate marketing/sales. For example, since my apps are made for entrepreneurs, many entrepreneurs need websites for their businesses. And it is a very natural thing for me to offer them services like website hosting. It is something they need, and appreciate it when I make a good recommendation. Since I don't provide website hosting as a service, I recommend other companies that do provide that, and collect a commission for the referral when a sale is made.

And at the same time, I get a commission and people get help. It is a win-win situation.

v. Making Money With In-App Purchases

Android and iOS platforms allow for in-app purchases. Pound for pound, the best in-app purchases are subscriptions because they aren't just one time transactions. If a user stays subscribed to your service for years, that user can ultimately bring thousands more percent in revenue to your business than if they only bought your product once.

The biggest challenge is making sure that whatever in-app purchase you offer makes sense for your users, and is something they really want or need. If it isn't something they want or need, they won't buy it. Think long and hard about what premium features your users might need, and make that into in-app purchases.

Also think about what features they might need on a regular basis. Regular-use features are perfect for selling subscriptions since the users use those features on an ongoing basis.

The next best thing to subscriptions is selling consumable in-app products. I'll touch on them shortly in an upcoming section that talks about the whale monetization pattern.

vi. Making Money By Selling Your Own Products

In the case of my apps, I sell my online courses, my books, additional apps, and a few other kinds of digital products. People don't mind it as much because I don't sell them via blatant banner ads. Plus, my books and courses teach people how to start and promote their businesses, so they are very relevant to what the users got my apps in the first place. If you have your own products that you can create and promote, it can be better than selling affiliate products because you keep a larger part of the commission and have more control over the quality of the user's experience with the product they buy from you.

vii. Making Money By Selling Services

In my apps, one of the things I sell is my business coaching. Since my apps are business apps, the apps are a way to show people that my apps and my help is valuable. People who feel that they get help from the apps, often want one on one coaching, and to be able to ask specific questions. I sell personalized coaching in two different ways. The first way I sell personalized coaching is as a premium feature. If people buy a paid $0.99 premium app, they can ask me business questions over chat right on the app. It is a very affordable option. For those who feel that they need much more help, I offer business coaching over Skype, which costs $50/hour.

If your app is some sort of a utility or business app, you can up-sell your services from the app. Keep in mind that the app store economy makes consumers want things for free or for very cheap prices. So it will be difficult to sell an expensive services. Nevertheless, it is a very reasonable option.

viii. The Whale Monetization Pattern

Most mobile app users will never pay for anything on your app no matter how much they want or need the extra features that come with the purchase. That is OK. It is just a part of doing business. Instead of focusing on those consumers, it can often be much more productive to identify and focus on the consumers that do pay. People that do pay are interesting because some of them may be able to spend a substantial amount of money with your business. The customers that spend a significant amount of money on your business are called whales because the amount they spend are whale-sized when compared to the rest of your customers.

Let's go over a popular pattern as an example of how some games use this whale pattern to extract as much money as possible from customers who are able to spend it. If you play mobile app games, you may realize that many of the games have an unlimited number of levels. There is no end to many of the games. You start out playing for free, and get hooked to the game because it is fun. People who get hooked eventually make it far enough in the game where they can't beat a certain level because the game just becomes too difficult. That is done on purpose by game developers. Once users are stuck on some level, they are given a choice to buy extra items or points to help them beat levels of the game for as little as $0.99.

Most people don't buy such items, but few people do. Once an individual has justified making such a purchase in their minds, making the same purchase again may not seem to them like a big deal. Guess what happens in the next level of the game. It doesn't get any easier. In fact, it gets harder and harder. To beat further levels these users have to spend $0.99 again and again.

Most people do this only a few times, and eventually stop. But some people who are hooked and addicted enough to the game have been known to spend thousands or even tens of thousands of dollars this way. These people are obviously rare, and their judgement and even emotional stability may be questionable, but they do exist.

These whale users tend to account for the bulk of the revenue for many games out there even though they make up a tiny fraction of all the game players for a given app. There is also an ethical question of whether this kind of an approach should be used since it takes advantage of people's psychological weaknesses who are maybe prone to addiction. Nevertheless, this pattern is one way in which many mobile app developers have found a lucrative way to make money from their apps.

For your part, as the entrepreneur, you must always think about ways to extract as much money from customers as you possibly can. It may not sound like the most ethical thing to do, but if you want your app or business to survive, your challenge is certainly to find paths to sustainable revenue sources.

And you should have no problem taking as much money as you can from customers as long as you give them amazing value in return.

ix. How To Maximize Your App's Revenue

Whatever your app may be, one of the most common strategies to maximize your revenue is to extend the relationship that your customers have with your business. The longer your users use your app, the longer they have to warm up to the idea of buying various items from you.

Try to think of features that can make the app something that your users use every day, and make a part of their daily habit. This isn't simple. Personally, I have been thinking for quite a long time about such features for my business apps, and while I've come up with features that satisfy this goal to some degree, honestly I am not happy with what I've come up with, and I am still thinking of something better. I feel that I have not found the perfect killer feature. I mention this not to complain, but rather to illustrate how challenging this can be. For some apps such features are natural, but for some apps it is extremely challenging. So if you don't come up with such a feature quickly, don't be discouraged. Keep it in the back of your mind and keep thinking about it.

Once you come up with a way to make your app a part of people's daily habits or tasks, it becomes natural to sell a subscription-based in-app purchase inside your app. The key to subscriptions is that once people subscribe, they rarely unsubscribe. It takes work to unsubscribe, and frankly, many of your users won't even be able to figure out how to unsubscribe. Think about how your local exercise gyms make money. They all work on the subscription model because many people exercise for a few months, quit exercising, but think that one day they will go back to exercising. So they keep their gym memberships. Because of this nuance, many business owners covet subscription-based revenue models.

If it doesn't make sense to create a subscription offer for the type of app that you have, there is another option that is nearly as good. You can simply create features that your users will want to come back to on a regular basis. That way you can get people hooked on using those features without making them sign up for a subscription. A subscription won't be as necessary if those people are naturally hooked to using a paid feature. The difference is that when those people stop being hooked, they will stop making the purchase whereas if they were signed up for a subscription, they would have to do a bit of extra work to unsubscribe, which is something many people wouldn't do.

Another way to make sure you put yourself in a good position to get your users to spend the maximum amount of money with your app is by offering multiple products. People who like your products are often open to buying more from your business. You just have to actually have something else to sell to them that they may want. In the case of my apps, they come as a 4-app series. 15-20% of the people who buy any of my paid apps go on to buy at least one more of my paid apps. That has been a very successful way for me to increase the average revenue from users of the apps who are open to spending money on my apps. In fact, you don't have to stop there. Some people go through all 4 of my paid apps, and want more. For those people I offer courses, books, and my coaching. It rarely happens that a person gets my apps, books, courses and coaching, but every once in awhile that does happen. When it does, that single person can bring hundreds or even thousands of dollars of revenue. They are my whales and I love them.

That concludes this chapter. Before moving on to the next chapter I want to take a little pause and see how you are doing. Are you enjoying the book? Is it helpful? Do you have questions?

You should always feel welcome to email me at: alex.genadinik@gmail.com

I would love to hear your thoughts about the book, and your suggestions for what I can potentially add to the book to try making it even better.

Want me to review your app?

If you email me with feedback, I will be happy to review your app in the app stores as well and hopefully give it a nice rating.

And, of course, if you are enjoying the book, it would really help me if you could leave a review of it on Amazon. Books really depend on readers like you to leave testimonials, and it would help me get the word out about this book if you took a minute to add a review on Amazon.

Chapter 4: Additional Tactics And Strategies

i. Comparing Mobile App Stores: GooglePlay vs. Apple App Store vs. Kindle vs. Windows Phone Store vs. Blackberry vs. NOOK

There are many app stores to choose from, and it is difficult to focus on all of them equally well. Let's go over the pros and cons of each app store when it comes to monetization, marketing, and development.

As I alluded to earlier, the Android platform is a good choice if you want to quietly release your app, and make many rapid improvements to it as called for by the Lean Start-up methodology of Eric Ries. If you are not familiar with the Lean Start-up methodology, it is a very widely adapted theory on how to rapidly improve your product by getting market feedback, and iteratively improve your product. Android also has the most global users, which is quite a nice advantage of releasing your app there first.

What Android's flexibility means for you is that you can release your app on GooglePlay, and then easily port that app to other Android-based app stores like the Kindle app store, the NOOK app store, and many other Android-based app stores in other countries. This can make for a bit of a management nightmare if you later have to make updates to all those apps, but you will be tapping into over a billion devices worldwide.

The Apple App Store, on the other hand, requires apps to be made only for it. Once you make a native iOS app, it can be sold only on the Apple platform for iPhones and iPads. You still get access to hundreds of millions of users, but the overall reach of Apple is now smaller than that of Android. Despite having a smaller reach, the iOS platform has a far higher number of affluent users who are willing to spend money on apps or in-app purchases. The amount of money spent per app user is far higher on iOS than on Android.

The Kindle platform is a good second option to reach US and European audiences with Android apps. The Kindle has a few interesting advantages. People who own a Kindle expect to pay for content on this device more readily than GooglePlay users because they expect to buy books for the Kindle. That is very different from other Android platforms where users go out of their way as much as possible to avoid paying. The other interesting thing about the Kindle is that you can sell books from your app as an affiliate. Since the Kindle platform is a very natural one for buying books, your conversion rate will be much higher for books than it would be on another Android platform like GooglePlay. The challenge with the Kindle is that its volume of devices and users is orders of magnitude lower than GooglePlay.

Since we just covered the Kindle app market, let's take a second to talk about Kindle's cousin, the NOOK device family from Barnes & Noble. While I am personally a fan of the NOOK platform, and have apps on it, it has been a beleaguered brand in the mobile app world with a very uncertain future. Despite me personally being a fan of the platform and wishing it the best, my advice would be to not invest in it by putting your apps on it unless you can port your apps very fast. The volume of downloads and the generated revenue will be very underwhelming for most apps. Plus, no one knows how long the NOOK platform will be around, and what will eventually become of it.

Lastly I want to add a few words about the Windows phone and even the Blackberry. They may seem like very different platforms, but in my opinion they share something very similar. In my opinion (and I could be wrong) they are both "too little and too late" to the game. Those platforms and their device install base are tiny compared to iOS and Android. Developing apps for those platforms has extra barriers associated with their unique technological platforms. The extra development barriers, low potential for distribution via those platforms, and the fact that they need to compete with some of the top companies in the world like Apple and Google, essentially spell doom for the Windows phone and the Blackberry. I personally don't recommend investing in your app growth on those platforms.

Android App stores based in other countries can be a good option in a few cases. There are countries with very large populations like China and a few others, which have their own Android app stores. The challenge with those countries is that even though there are many potential users for your apps there, those users will on average be far less lucrative than users from economically strong countries like United States, Australia, Canada, UK, and the rest of Europe. Plus, many people in developing countries don't use mobile banking which means that even if they wanted to buy your in-app purchases, logistically, they can't. That means that a large part of the decision of whether to put your app on app stores in those countries depends on how your app makes money. If you make money from ads, despite generating much less revenue in economically poor countries, they will still generate a small trickle of revenue. If you make money with in-app purchases on the other hand, the drop in revenue will likely be far steeper.

ii. Should You Develop Your Apps Natively?

There is a debate in the mobile app world regarding whether app developers should develop their apps natively or use technologies that enable a single code base to be compiled into both, an iOS app and an Android app. There are a few technologies that enable developing for both an iOS and Android at the same time. Some popular examples of these so called frameworks are Appcelerator, Titanium, and PhoneGap.

The factors that go into the decision of developing mobile apps natively or using such frameworks are relatively straightforward. Android and iOS apps are different in their feel and usability. Creating 100% identical apps for both platforms doesn't make the best sense in many cases. Nevertheless, it is obviously much more expensive (in money, skilled labor, and time) to develop unique apps that are native to iOS and Android.

So if you are OK with your apps looking only 80-90% natural to the platforms they are on, using a framework that will take one code base and compile it into apps intended for multiple platforms may be a great way to go. This way you can easily also create a Blackberry app or whatever other platforms it is possible to compile your app into.

On the other hand, if you develop natively, it will be more possible to get to optimum usability and quality for each of the platforms on which you plan to distribute your app.

To give an example case study, in the case of my apps, I decided to develop everything natively. That meant creating a whole new app for iOS and a whole new app on Android. I had to use different programming languages and different technologies that enabled those programming languages. That was difficult and it took a substantial amount of time because despite me having years of prior software development experience, there were many things I had to learn from scratch. But developing natively gave me one advantage that was key. I didn't need to make identical app updates to both apps at the same time. I sped ahead on Android with constant experiments to get the app to generate more money, get more users, and increase user engagement. Most of those experiments failed, but a few of those experiments worked. The experiments that worked ended up eventually making their way into my iOS apps. So despite having to do much more work, developing natively on each platform gave me flexibility in what I was able to do with the apps.

Overall in the mobile app industry, most companies that can afford it, develop their apps natively. Most of their app updates loosely mirror each other for iOS and Android, and they prefer having the flexibility to do what they want on each platform. Most companies also prefer being able to make each app look as close as possible to what it should be on any given platform.

If you need to develop an app quickly, or simply don't have the resources (time and money) to develop natively, then using a framework like PhoneGap, Appcelerator or Titanium begins to make much more sense.

iii. General Features To Add To Your App That Will Help It Grow And Make Money

Let's have fun and brainstorm some potential features of your app, no matter what your app might be, which can help you get more downloads, make more money, get better reviews, and increase user engagement. Think about how the features covered below can be added to your app.

As I always allude to, if you can add some sort of chat where users either chat with you or each other, it will create a natural pull for users to get back into your app and to keep opening it. If you can engage with some of the users of your app, especially during the early stages of your app, they can explain to you what they like about your app and where your app falls short. There is nothing like getting real feedback from actual users. There is one key to success with this strategy. The things about which you chat with your users must be interesting or helpful to them in some way. If it isn't, the users won't bother to talk to anyone. The chat-based help or advice can become one of the paid features as demand grows. As an example, consider how cool it would be to get expert help on a dating app to help premium members create better profiles and attract more people.

In the theme of helping your users with something, if your app can educate your users about the topic they are interested in (not common for game apps, but often possible in other niches), you can add written or video content to your app. The benefit to this is twofold. The first benefit is that your users will appreciate your app more, and may possibly leave better reviews. The second benefit is that when your users engage with those educational materials, they are increasing the your app's average sessions and session lengths, which helps your app rank better in search. Plus, if you happen to create premium content regularly, that could be a subscription opportunity.

Another good feature to add to your app is something that would get your existing users to invite their friends to use some features of your app together. This can often work well for games because playing with friends is more fun. But this can work well almost no matter what kind of app you have. You can incentivize users to invite their friends in other ways by giving them access to extra features, credits, or anything else that isn't totally free in your app.

Just as you want to think about features that will make your app better when people use them with friends, you also want to come up with features that will be something your users will want to use on a daily basis. If the app can become a part of people's daily habits, they will be far more likely to warm up to the idea of spending money to buy premium features and tell their friends about your app. Think of what people do every day that may be natural for your app to help them with. If you come up with such a feature, it may be the "killer feature" that helps to skyrocket your app's engagement and ultimately search ranking, growth, and monetization.

If it doesn't require too much extra work and development time, consider adding gamification features. Competing on points or status with other app users makes using the app fun and more addictive. Just don't necessarily make gamification a part of your initial features as you launch because at times, gamification features can be complex to create. If you make your app more complex to create, it will cost more and take longer to develop and launch, increasing the risk that it never actually gets launched.

iv. What Kinds Of Apps Get An Investment?

Almost all entrepreneurs, at one point or another during the life-cycle of their business, wonder about raising money for their business, and take some steps to get that money. For apps, the popular fundraising strategies have been crowdfunding, investors, and friends and family.

For now, let's focus on investors and understand what kinds of apps or businesses have potential to get an investment, and what kinds of apps don't. Before I start to get into details of what is possible and what is not possible, I want to say that every investor is obviously different. They are usually very smart business people, but they are still people. They can often be wrong. There have been many cases where one investor told an entrepreneur that they have a terrible business, and a different investor invested in that business, and the business went on to be a major success.

It is also possible that neither of those investors was wrong. Perhaps the investor who declined didn't like the risk-reward situation and the ladder investor was more risk tolerant, or liked the entrepreneur more than the business, and after much hard work they were able to create a great business out of the app that once didn't seem like it would be a great business at all.

Despite the unpredictability and variation of different kinds of apps and different kinds of investors, there are a few guidelines that investors like to use, and will likely use to evaluate your app.

Pound for pound the thing investors prefer most is that your app targets a large (billion dollar plus) market. This ensures investors that your app can potentially grow into a business that can give them returns that are large enough to justify their risk. For this reason, investors seem to invest in what often appear to be similar kinds of apps: photo apps, dating apps, health apps, social apps, fashion apps, commerce apps, sometimes game apps, and a few other kinds of apps.

Game apps are a bit of a wildcard. Think about game apps. Some games are duds and some games are winners. Even great game development companies can't consistently produce great hits. The game industry is sometimes called "a hits industry" meaning that most money is made from games that become huge hits. But creating games that are hits is extremely difficult, and not something that game developers can easily predict. It is extremely difficult to predict game success. Even if a game seems to be successful, the popularity of many games quickly vanes, and there is no certainty that the developer of the game will ever create another hit. That adds unpredictability and extra risk to game development companies in the eyes of investors. Plus, if you are a single developer, it is incredibly difficult for you to compete with big game development studios in marketing and app quality. That makes it difficult to get traction for games, and lack of traction makes it difficult to get investors interested.

Another thing investors really like to see is high growth. Growth is important for all start-ups, but for apps it holds even more importance because for the most part, apps that win are apps that manage to become very well known and very widely downloaded. This can be understood by looking at the business models of successful apps. Most apps do not generate tremendous revenue on their own, and therefore do not become great businesses on their own. The great success for apps is to get acquired by a larger company. I realize that this is the case for most businesses, but for apps this dynamic is just more glaring.

To get acquired, apps need serious growth. When I say serious growth, what I have in mind is hundreds of thousand or millions of downloads and/or upwards of 10% month over month growth on a consistent basis. If you just got disappointed because your app can't boast that kind of growth just yet, or has not been launched yet, I am sorry to be the bearer of bad news. But with enough hard work and if you correctly implement the strategies in this book, I am sure you will get there. I hope you do, and I am rooting for you!

When thinking about the potential of getting an investor, it is important to understand that investors aren't there for your sake. They are there to capitalize on your success or potential to succeed. If you have not launched your app yet, there are many apps out there that have launched, are growing rapidly, and are more compelling to investors at that moment. But if you work hard and achieve growth, you increase your chances to eventually get an investment. Just focus on growing your business without getting an investment if you don't immediately get an investment. Most businesses never get an investment, and it isn't something you can rely on or wait for. I never got an investment. I had to scratch and claw, and so do most entrepreneurs who eventually become successful.

Lastly, I want to mention another thing that investors look for before they decide to invest in an app or not. That factor is your team. Many investors frown on single-member start-ups, which many app companies are. In fact, my own app company is a single person company, and whenever I talked to investors about my apps, they immediately asked why I don't have co-founders and why I am not growing my team.

An ideal mobile app founding team, or any start-up team for that matter, is one in which the founders have successfully worked together in the past, have experience in the business niche in which the business is in, and have a balanced skill set.

v. Crowdfunding For Apps

Since most mobile app entrepreneurs are unfortunately not able to raise money from investors (if you think about how many apps there are out there and how many have gotten an investment, far less than 0.1% of apps ever get an investment), one of the next best options is to raise money by getting donations via crowdfunding.

There are a number of crowdfunding sites out there like KickStarter.com, IndieGoGo.com or GoFundMe.com. Those are general crowdfunding sites for many kinds of projects, and you would be in relatively good shape trying to raise money via those platforms. But there is one crowdfunding site that focuses specifically on raising donations for mobile apps. That site is AppStori.com.

[Revision: AppStori.com has gone out of business since I originally wrote this book in early 2014. The latest revision is December 2015. I leave mention of it here as an example of just one of the businesses that found it challenging to succeed in the mobile app world.]

I realize that still not everyone knows exactly what crowdfunding really is so let me briefly explain this concept to you. Crowdfunding is a way to get donations online from strangers, friends, and family who believe in your project, or simply want to support you. You get many small donations from many people across the Internet by leveraging the gigantic scale of the Internet.

That might seem great, and it can be. But there are many caveats to crowdfunding. First of all, because this is essentially free money, entrepreneurs flock to this fundraising strategy creating an overly saturated environment. Due to the over-saturation of companies that are trying to get funding via crowdfunding, this has become far from the "free money" many people see it as.

For your crowdfunding campaign to do well, you can't just sit back and watch the cash come in. You must promote your crowdfunding page, and get your friends, family, business contacts and people who you reach with your marketing efforts to donate to your crowdfunding campaign. Arguably the worst part of it all is that on most of these crowdfunding platforms, you don't get any of the money you've raised unless you reach your stated goal which you declare in the beginning of your campaign. The reason for this is that the crowdfunding sites realize that you will just sit back and watch money come in unless you are afraid of not getting any of the money. If you are scared of not getting any of the money you've raised, you will be promoting your crowdfunding campaign with all of your might. And guess what, when you promote your campaign, you indirectly promote those crowdfunding sites because the campaigns are hosted on their sites, and they take a cut from the money you've raised.

To sum up, you have three options when it comes to raising money for your app through crowdfunding. The first is to sit back and hope that people will just find your project and donate to it. This doesn't work unless you are able to somehow build momentum through your own efforts. Once projects get momentum within these crowdfunding sites, they tend to float up on the crowdfunding sites and naturally get more exposure. But you have to push for that. The second option is to ask your friends and family for donations. Frankly, my opinion is that in most cases, if you do that, you might as well bypass crowdfunding sites, and just ask people you know for money directly, saving yourself commission from those sites. Lastly, if you have access to a strong marketing channel where you know that you can reach many people, promote your fundraising campaign there, and hope that people will donate.

vi. 10 Fundraising Strategies

This section is a quick summary of the ten fundraising strategies listed in my full fundraising book. For a full discussion of these fundraising ideas, consider getting my full fundraising course as one of my free gifts to you that I offer at the end of this book.

Let's quickly go over the possible strategies or ideas you can use to raise money for your apps. Keep in mind that these are strategies for general ways to raise money, and not specifically for apps.

1. Investors - as we covered earlier, it is very difficult to raise money by getting an investment. For a full list of professional technology investors, explore angellist.com and see if any of the investors there invest in the kinds of apps that you are working on.

2. Donations via crowdfunding - as we covered earlier, this is a viable but difficult way to raise money. In most cases, to make it work, you will have to be the one who will have to reach out to many people asking for the donations.

3. Loans - while getting a loan isn't recommended because it is an extra risky strategy (potentially losing money you don't actually have), you can get a small loan (sometimes called a micro-loan) on sites like Lendio.com or Prosper.com or other micro-loan sites. Banks don't usually give business loans to businesses which have not been started or businesses that have minimal revenue, so the micro-loan sites are the only viable option for idea-stage and early-stage businesses.

4. Grants - there are very few grants for apps specifically. Nevertheless, you can apply for local grants, minority grants, and any other grants for which you may be a fit. Explore grants.gov for a full list of possible grants.

5. Getting part-time or full-time work - I realize that this does not seem appealing to entrepreneurs, but it is a great way to

ensure that you can have money coming in, and putting some of that money towards your app. I encourage people to embrace this as an option because it is least risky in terms of getting at least some money in return for the efforts you put into your fundraising.

6. Generate revenue from your business - this is another "not so popular" suggestion, but it is one that works. If you can, try to explore possible revenue streams early on in the lifetime of your app. Monetizing your app can slow the growth of your app, but if you need money to sustain your business, this may be a very realistic option.

To be honest with you, tips 5 and 6 are what I used to raise money for my business.

7. Creative ways to raise money - consider putting on some fun events, fundraisers, or something with your local community to help you raise money towards your business. In my fundraising course I give an example of how I created an event series around one of the businesses I was trying to grow. The events helped me promote the business and generate revenue at the same time.

8. Raise money by selling to future customers - one very savvy strategy for some businesses is to approach future potential customers and try to sell them products or services for a very heavy discount in exchange for upfront payment that you can use to actually develop the product. This helps in many ways. It obviously generates revenue, but it also gives you practice selling your products, and helps you determine

how excited your potential customers are about the product you are building.

9. Provide educational services, materials or workshops. My own business is a perfect example of this. I created courses, books, a YouTube channel, and a number of other educational materials which help my current app users, help new customers find my business by discovering that educational content, and generate revenue that I can put back into the business. Think about what educational materials you can create that make sense for your business niche, which you can sell to attract your potential customers and generate revenue at the same time.

10. Provide services online - if you need to generate cash, consider freelancing on sites like elance.com, fiverr.com, odesk.com, fancyhands.com, or any other freelance or concierge site.

vii. Dive Into Mobile App Business Models

Understanding mobile app business models a very powerful tool to help you during the planning stages of your mobile app to help you understand what kinds of apps may find success, and which will fail. It will also help you evaluate potential app features that you may consider adding to your app once your app is live. I can't overemphasize how important this is to the health of your app business, and I hope you agree with me as we cover mobile app business models. One caveat I want to add is that because there are so many different kinds of apps, I am not able to cover specifics of different kinds of apps, and at times I have to make sweeping statements and assumptions. Please bare with me through that, and try to consider how the general points I make relate to the specifics of your app.

Before delving into business models, let's first define what a business model is so that we are all on the same page, and are working from the same definition. Besides the many different accurate definitions of varying complexity for what a business model really is, this term is often used inaccurately, adding to the confusion.

Here, I'll try to present a simple but effective definition of a business model:

The business model is a bird's eye view of your entire business, and how your business functions. It is a way to step back and take a look at each and every component of your business individually, and evaluate how well each individual component of your business works on its own and together with the other components of your business.

As you may guess, a business has many components, and things can quickly get quite complex. Luckily, mobile apps are relatively simple businesses to examine as a whole because many traditional components of a business simply don't exist for apps. When it comes to apps, there is no shipping, handling of inventory, no logistics, no manufacturing of physical goods, often no rent to pay, and there are typically fewer employees needed than most traditional businesses.

For that reason, when it comes to apps, we can simplify our understanding of what is a business model. Let's take another step back, and explore the three core components of any business, and let that be our business model evaluation for now. This way of looking at things will get us to 90% of where we need to be in terms of understanding our business model without many of the unnecessary details.

The three core components of any business model are: product, marketing and finances. Let's cover these in more detail.

The first component of your business model is your product. Within this category, we should consider the resources it will take to create, maintain and continuously improve your app (this is your product), and what it would take for the product to be competitive in its business niche.

The second component of the business model is marketing. You must consider the type of app your company offers, and evaluate how you will promote that unique app. You must consider the natural marketing strategies for your app, and be able to estimate the kind of distribution potential your app has, and how attainable it is to reach the maximum potential with each of those natural marketing channels.

Lastly, you must tie all this together by evaluating the financial picture of creating your product, selling it at the right price and volume to cover your expenses, break even financially, and eventually reach your financial goals for your business.

Once you come up with a realistic plan which allows for the creation of a good product, ability to promote and sell that product, with all that being done profitably, you have a foundation for a winning business model.

Now let's take a look at an average mobile app business model by looking at mobile app business models in this light. You may recognize that the biggest themes in creating a mobile app business is precisely the creation of the app, the promotion of the app, and the monetization of the app. **We have essentially been discussing the business model components all along, without explicitly stating so.**

Let's start with the development of the app. If will be far cheaper to develop the app if you can build it on your own rather than having to hire someone. App development is the single biggest cost for just about all apps, and it substantially changes the financial dynamics of creating your app compared to if you can create the app on your own. If you can't develop the app on your own by having at least one software developer on your founding team (which could be you), your business model automatically becomes far weaker. This is a very serious issue to take into account and not gloss over.

Marketing of your app, which is the next part of your business model, is not as black and white. Recall how the bulk of downloads for most apps comes from apps store search, and the next sources of downloads come from social sharing and publicity. First, consider how social/viral your app realistically is, and how much big press coverage your app will be able to get. Next consider how well you will be able to rank for your most coveted search keywords in app store searches, and how much approximate volume of downloads those searches will generate. Everything mentioned in this paragraph is an educated guess during the planning phases of your app. Complete accuracy is not possible, but if you are realistic about the distribution potential for your app, you will come close to a reasonable estimate of the daily or monthly download potential for your app. At least you should do your best to come close to realistic numbers as possible.

Once you have an estimate for your app's download potential, the focus should shift to the monetization potential for your app. While it is difficult to cover monetization for every possible different app in this single paragraph, by and large, mobile app monetization tends to be poor because users don't like paying for apps or items within apps, and if you promote other things (ads or products) too aggressively within your apps, you will begin to have bad reviews which will work to undermine future downloads and monetization.

Now let's tie all of this together into a succinct view on mobile app business models. Since generally, monetization is poor, you often need a substantial number of downloads for your apps to break even or pay yourself and all the people working on your app a reasonable monthly salary. And since you can estimate your app's approximate download potential, consider whether downloads from search, social sharing, and publicity can amount to tens of thousands of monthly downloads, which is typically the minimum number of downloads you need for a free app to make sufficient money to begin covering your salary and the salaries of anyone else working on the app with you.

As you can see, things do not seem favorable largely because of poor app monetization even if you can generate tens of thousands of monthly downloads. Your app will begin to turn into a business only after you get hundreds of thousands of monthly downloads, or you figure out a very strong monetization strategy.

So what are the takeaways? To me they are clear. If you have a unique strategy to consistently generate tremendous monthly downloads, or a very strong monetization strategy that will monetize at a far better rate than an average app, the app idea seems worth pursuing. If you don't have such a strategy, it may be a sign that the business model for your app is not as strong as is necessitated by the current mobile app ecosystem.

viii. Mobile App Exit Strategies

We make our apps in order to achieve some kind of a goal, right? Some people want their apps to be the tools by which they become millionaires or billionaires. Other people just want to earn a comfortable living and generate passive income, and yet others simply want to achieve a career boost and some kind of technology-world notoriety. When you plan your app, you must set your goals for it. Sometimes these goals can be synonymous with the term exit strategy. In general, there are three kinds of exit strategies (common outcomes) for an app that can be considered successful. Let's cover them in some detail.

GETTING BOUGHT FOR A SIGNIFICANT AMOUNT OF MONEY

The first and most appealing exit strategy to most people is to sell their apps for a large sum of money to some larger company. I personally, would be very happy if a larger company contacted me, and wanted to buy my apps for millions of dollars. Unfortunately, it isn't that easy.

To achieve million dollar buyouts and notoriety, an app must generate tremendous scale. Tremendous scale here means millions or tens of millions of downloads or even more. With those kinds of numbers your app will stand out in the app stores, you will be able to get publicity, and potential larger companies who may eventually buy your app will learn about your app from your publicity and prominence in the app stores. The challenge is that to achieve that kind of scale, an app must be truly extraordinary, have a significant marketing budget, or do something very unique to grow.

Apps that eventually get bought and have lucrative exits tend to have one thing in common. They catch on with the general public, and often become household names. This is a level of cultural penetration that is far beyond just dominating search or getting publicity. To achieve this, the app must be life altering or the word of mouth and social invites of the app must be absolutely through the roof. Without an extremely wide adoption, regular use, and incredible engagement, it is nearly impossible to achieve such a status.

SELL THE APP FOR JUST A FEW THOUSAND DOLLARS

If you launched your app, and are having a hard time growing it to incredible scale, you can sell the app to another mobile app entrepreneur.

A website like flippa.com is a marketplace for apps that are currently for sale. You can either sell your apps or buy existing apps for a very affordable price.

DON'T SELL YOUR APP AND CONTINUE TO GENERATE REVENUE FROM IT

If you find it impossible to generate tens of millions of downloads, and you don't like the price that you can get for your app on flippa.com, you can simply keep the app and have it generate revenue for you. How much revenue is up to the kind of app you have, the number of downloads you can generate, and other nuances specific to your app. For most apps, revenue isn't too big because as we covered before, mobile app users don't like to pay, and ads don't generate much revenue at all. But the good news is that you can have a steady income from an app for a long time without putting more effort into it.

iv. How To Get Featured In App Stores

Many app entrepreneurs ask me about getting their apps featured in app stores. This is very difficult to achieve. My Android business planning app was once featured in the Google Play app store, and I still can't tell you what the exact formula is to guarantee being featured. I can only share my observations with you on what I think I did right for my apps, and what I've seen other apps do that I think increased their chances of getting featured. Let's cover some of these strategies.

First I'll give the case study of how my own app was featured since I had this experience first hand, and can share details. I wasn't told about this by the GooglePlay app store. One day I simply woke up and I had a spike in downloads by about 500 downloads per day. Surprisingly, this lasted for a few weeks (I would have thought that it would have been over sooner), and for some reason, most of the downloads came from England. Possibly my app was featured in additional ways there, but I had no way to confirm that since I was based in United States, and couldn't see what an app store users in England would see.

Since the app store didn't inform me of anything, I had to figure out why I was being featured. I tried to understand the reason for being featured, and the only reason that made sense to me was that a short time prior to being featured, I had added new app icons and improved my app landing page, which boosted my download totals by 15%. The increased download totals helped to boost my app's search rankings, which gave me an even bigger daily download boost. My conclusion was that the download boosts made my app stand out, and the app store editors noticed my app out of the many others, and decided to feature it because the app was already showing signs of being liked by people.

This kind of a pattern where boosts in downloads result in being featured is something I see frequently. There is no guarantee that a boost in downloads will get an app featured, but the apps that do get featured seem to have this in common in one way or another.

When apps of my fellow entrepreneurs got featured in app stores, it was a similar story. First, the app had to be good enough to garner some sort of attention. Then their apps would get serious publicity or the app entrepreneurs had to generate spikes by paying for ads, or by taking advantage of app store search (like my app). Once that happened, strong spikes in downloads had very strong correlations with eventually getting the app noticed by app store editors, and getting featured.

I won't mention which apps I know that got featured because of what I will say about these apps now. Except for my apps (I am an independent developer), all these apps were built by wealthy companies that got millions of dollars in funding. With that kind of funding, many options open up that aren't too available to mere mortals. The companies that made these apps were able to buy publicity and get stories about them published in Mashable, Techcrunch, and even get television appearances. All of that obviously drove download numbers through the roof which got the notice of app store editors, and eventually got the app featured.

Additionally, because these wealthy app companies had investors, they often leveraged the relationships of their investors to be able to contact the app store editors themselves, and to be able to influence them directly. Unfortunately, few app developers have that opportunity, so we just have to be that much more scrappy and resourceful.

Chapter 5: Common Types Of Apps And Strategies For Them

i. Game Apps

Many mobile app developers dream of making the next cool game. In fact, many developers start developing in the first place because they want to make games. I am not a mobile game developer myself, but in my coaching practice I have worked with a number of mobile game developers. With each app, we consistently run into the same few issues, and I want to outline some of those challenges here so you are aware of them, and have a plan to get around them.

The first challenge for game apps is generating large numbers of downloads. As we covered earlier, the main drivers of downloads for most apps are app store search, publicity you may be able to get for your apps, and existing users inviting their friends to use your app. While there are many other strategies to promote apps, these are some of the top yielding strategies for most apps.

Since app store search is so important, let's start by considering some nuances of app store search for game apps. The problem for most games is that there is a very limited number of keywords that games can rank for. Those keywords repeat across almost all game apps. Some examples of such keywords are: fun, run, jump, game, puzzle, animal, adventure, shooting, race, points, kids, etc. You can see how most games can try to rank for a number of these keywords. The problem, of course, is that there are tens of thousands or even hundreds of thousands of games that any new game app would have to compete with when it comes to those keywords. And the top search spots are already taken by very strong and popular apps that have large marketing budgets, access to press, and are often featured by the app stores. It is nearly impossible for new apps to compete with them in app store search.

Since app store search is obviously a very difficult option, the potential for social invites is also severely reduced because social sharing requires an initial install base of people who like your app and share it. Since it will be very difficult to get a large number of people to install your app from app store search, social sharing potential will also be limited.

To get publicity, your app has to be truly extraordinary. If this was year 2008 or 2010, it would be much easier to get publicity for apps. The problem is that new apps aren't exciting these days unless they are truly groundbreaking, have serious money or people behind them, or have a truly unique story that people either fall in love with, or is something that makes them curious.

To generate downloads, you can pay different app review sites, or explore other marketing channels. Chances are that paying for app downloads will happen at a loss, and the revenue from the app won't cover the marketing expenses. Other marketing channels, for the most part, require a lot of effort and bring only mediocre numbers of additional downloads.

If this feels depressing, it is. At least it is for me because I work with many people who face this kind of a situation often. I want them to succeed, but there is a limited number of solutions I can offer if the problems they come to me with are this challenging.

So before you begin creating your game, try to come up with a unique or interesting enough story that will help your app stand out from the mass of existing apps out there. Being good isn't good enough these days when it comes to games. The game must be truly something special or something groundbreaking in some industry. Ask yourself whether people will truly love the app, and how you can get a cult following built around your app. Think of ways to make your game special, unique, and fun. The bar of quality and user expectation is very high.

Consider these issues during the business planning stages for your app rather than after the app has been developed. Thinking through solutions for the issues raised here before you start can save you a lot of money and headaches down the line.

Before moving on from game apps I want to suggest an intermediate to advanced ASO strategy that might help your game app rank in app stores. If your game is a puzzle game or a game where your keywords would be very competitive ones like run, shoot or jump, my advice is to try to find less competitive keywords to try to rank for first. At least this way you will get a trickle of downloads instead of none. Use that trickle of downloads to make your app stronger in app store search. Once your app becomes more competitive in search, you can slowly start to target more and more competitive search keywords. This is actually what I did for my business apps, just with business related keywords.

ii. Social And Photo Apps

We just went over how to promote game apps. Unfortunately, many of the same challenges come up when it comes to creating photo apps or social apps. I am grouping them together because these are the kinds of apps young entrepreneurs love to build, and spend time and money developing. They often come up with cool apps, but have an extremely difficult time generating downloads for their apps. And if you can't get customers for your business, that can destroy the entire business.

Just like for game apps it is extremely difficult to rank for terms like fun, run, shoot, adventure, and similar other terms, photo and social app entrepreneurs find it equally difficult to rank for terms like photos, social, and friends.

The same pep talk takes place when it comes to these apps as with game apps. If this was the year 2010, you would be able to get a great amount of downloads, but today such apps must be truly groundbreaking to get publicity and stand out.

One aspect of social and photo apps that is more difficult than game apps is monetization. Games have much more clear monetization patterns than social or photo apps do. Social and photo apps typically have to come up with some unique in-app purchase offers. The social or photo app entrepreneurs have to not only figure out how to get downloads (a big challenge), but they also have to experiment with monetization strategies to find one that converts users at a good rate.

I can give examples of other types of apps with similar issues, but I think at this point you are getting the idea. There are new apps in the app market every day, and new ways to discover apps aren't appearing nearly as fast as the progress of app saturation. It has been like that since 2008. At this point, realistically speaking, some niches are nearly impossible to compete in unless you have some preexisting large marketing channels or are willing to spend the time to create one. One app store strategy I recommend, and one that I have myself used with success is to find niches where you can either make a fantastic app, or that are only semi-competitive.

Another savvy app store ranking strategy can be to target search terms that are easy to rank for when your app is new. Those terms won't bring you many downloads, but at least they will bring you a small trickle of daily downloads. You can use those downloads to increase your app's strength in search by making sure that the social and engagement signals of your app are positive, and are better than the social and engagement signals of your competitors. And as your app slowly rises in the rankings and becomes stronger in search compared to your competitors, you can adjust your title and description to attempt ranking for increasingly more competitive keywords.

iii. Dating Apps

Dating apps are another popular kind of app that app entrepreneurs like to create. It is also an attractive kind of app for investors as well. But by now, you know me. I like to play devil's advocate, and point out the challenges for the app so that you can be aware of the challenges and have a plan to overcome them. If you can figure out how to get around the challenges during the planning phases of the app, you will be in far better shape than if you had to think about how to get around those challenges after meeting those challenges head on after you've started.

As you may already guess, with dating apps, just like with game apps or social apps, ranking in app store search is one of the biggest difficulties. Monetization is also very challenging because more and more dating apps and websites are free.

Unfortunately, I don't have a solution for you for these challenges. What is even worse is that I actually have an additional challenge. For dating apps and websites to be effective, you must have what is called critical mass where enough potential daters are using your app at the same time within the same geographic area to ensure that they meet enough potential mates. Imagine what happens when one potential dater begins using your app. They immediately begin searching for potential mates in their local area. There must be no shortage of potential daters because a person needs to see many people before they like someone and contacts them. Considering how picky people are with whom they choose to date, you must have geographic areas very saturated with users in order to achieve a critical mass.

Chapter 6: Case Study Of ASO For My Apps

Let's now go over how I did ASO for my app which has ranked #1 for the term "business" for 2 years and "business plan" for 3 years, and who knows how long it will continue to rank for those terms.

Here is that app. You can see the title and description for yourself. And feel welcome to download the app. It is free:

https://play.google.com/store/apps/details?id=com.problemio&hl=en

If for some reason you are not able to visit the URL at the moment of reading, here is the title of the app:

Business Plan & Start Startup

You might immediately notice that my brand isn't in the title and that the title doesn't make complete English language sense. But you might also notice that I target many ASO keywords like:

- business
- business plan
- start business
- startup
- start startup

Now let's take a look at the description:

Top business plan app with 100,000+ downloads and over 23,000+ people who already created their business plans on this app. The app helps you in these ways. You can:

- Create your business plan privately or with friends and business partners.

- Walk through tutorials for every section of a typical business plan.

- Get a step by step timeline that takes you from planning to starting your business.

- Download your business plan to email for free.

- Get business planning help for your startup from our community.

- Get organized and motivated by maintaining a to-do list.

- Coming soon: network with entrepreneurs in your area, more guides and small business help and tutorials.

- YouTube videos and podcasts on topics of startups, and creating a small business,

In addition to the business planning and motivation tools, you can ask questions to our entrepreneur community about any part of your business plan or any part of starting a business, all right inside this mobile app.

The app helps you in 3 distinct ways:

1) By teaching you how to think about each section of a business plan, and empower you to not only create a great business plan, but also start your small business or startup the best way possible.

2) Software tools to help you create a small business plan, stay on track and motivated, and plan your business with friends and co-founders

3) Ask questions about your business plan to our entrepreneur community and get help from your fellow small business owners. And if you want expert help, please get our premium app where we offer professional help with your business plan.

We wish you good luck, much success and enjoyment from your business.

For our privacy policy and the rest of our mobile apps, please visit http://www.problemio.com

For tech support or other business help, please email the app developer: alex.genadinik@gmail.com

Notice that I reiterate many of the keywords I am trying to rank for in the description, and all the long-tail and smaller keywords that I target in the description.

Since I target so many different keywords within my niche, they all bring me downloads. And from that, I have a higher download rate than other apps that are competing with my app in search on any of the keywords that my app targets.

Do you see the trick here? By ranking for as many different keywords that I could, I am getting more downloads than my competitors, and using that as one of the advantages to continuously outrank them.

I do want to caution you. What I am doing with my app's title and description is a bit extreme. This approach isn't for everyone, but I have been able to get it to work for me, and I hope that this example helps your app do the same!

THE END! THANK YOU FOR READING

If you are enjoying the book, it would really help me if you could leave a review of this book on Amazon. Books really depend on readers like you to leave testimonials, and it would help me get the word out about this book if you took a minute to add a review on Amazon.

FREE GIFTS AND DISCOUNTS FOR YOU

Gift 1: **I will give you one free online business/marketing course of YOUR choosing.**

I teach over 100 online courses on business and marketing. As a special offer for readers of this book, I will give you one course for absolutely free, and you get to choose which one. Browse my full list of courses and email me telling me which course you want, and I will send you a free coupon to get free access to the course!

Here is my full list of courses to choose from:

https://www.udemy.com/user/alexgenadinik/

Just send me an email to alex.genadinik@gmail.com and tell me that you got this book, and which of my courses you would like for free, and I will send you a coupon code to get that course for free.

Want more than one course? To get additional courses for free, when you email me, just tell me which other courses you would want for free, and I'll see if I have any ongoing free promotions for those additional courses.

Gift 2: Free business advice

If you have questions or want feedback about your apps, your overall business, or anything mentioned in this book, email me at alex.genadinik@gmail.com and I will be happy to help you.

When you email me please keep two things in mind:

1) Remind me that you got this book and that you are not just a random person on the Internet.
2) Please make the questions clear and short. I love to help, but I am often overwhelmed with the amount of email I get, and always short on the time that I have available. So if you keep your questions clear and brisk, I will be able to answer them more quickly.

Gift: 3: Get my Android and iPhone business apps for free.

My apps come as a free 4-app course and iPhone and Android.

Free business plan app:
https://play.google.com/store/apps/details?id=com.problemio&hl=en

Free marketing app:
https://play.google.com/store/apps/details?id=com.marketing&hl=en

Free app on fundraising and making money:
https://play.google.com/store/apps/details?id=make.money&hl=en

Free business idea app:
https://play.google.com/store/apps/details?id=business.ideas&hl=en

Here are my free apps for the iPhone:

Free business plan app:
https://itunes.apple.com/us/app/business-plan-and-coach/id554845193

Free marketing app:
https://itunes.apple.com/us/app/marketing-advertising-articles/id587238156?ls=1&mt=8

Free app on fundraising and making money:
https://itunes.apple.com/us/app/funding-fundraising-ideas/id624657810?ls=1&mt=8

Free business idea app:
https://itunes.apple.com/us/app/small-business-ideas-help/id583498069?ls=1&mt=8

COMPLETE LIST OF MY BOOKS

If you enjoyed this book, check out my Amazon author page to see the full list of my books:

http://www.amazon.com/Alex-Genadinik/e/B00I114WEU

And here is my website with all my work:

http://www.problemio.com

DID YOU ENJOY THE BOOK?

If you liked the book, I would sincerely appreciate it if you left a review about your experience on Amazon.

Thank you for reading and please keep in touch!

ABOUT THE AUTHOR

Alex Genadinik is a serial entrepreneur, software engineer, and a marketer. Alex is the creator of the Problemio.com business apps which are some of the top mobile apps for planning and starting a business with 1,000,000+ downloads across iOS, Android and Kindle. Alex has a B.S in Computer Science from San Jose State University.

46185829R00070

Made in the USA
San Bernardino, CA
01 March 2017